SpringerBriefs in Neuroscience

For further volumes:
http://www.springer.com/series/8878

Kenneth Blum · John Femino
Scott Teitelbaum · John Giordano
Marlene Oscar-Berman
Mark Gold

Molecular Neurobiology of Addiction Recovery

The 12 Steps Program and Fellowship

Kenneth Blum
Scott Teitelbaum
Mark Gold
School of Medicine
McKnight Brain Institute,
University of Florida
Gainesville, NY
USA

John Femino
Meadows Edge Recovery Center
North Kingstown, RI
USA

John Giordano
Department of Holistic Medicine
G&G Holistic Health Care Services, LLC.
North Miami Beach, FL
USA

Marlene Oscar-Berman
Neurology, Anatomy and Neurobiology
Boston University School of Medicine
Boston, MA
USA

ISSN 2191-558X ISSN 2191-5598 (electronic)
ISBN 978-1-4614-7229-2 ISBN 978-1-4614-7230-8 (eBook)
DOI 10.1007/978-1-4614-7230-8
Springer New York Heidelberg Dordrecht London

Library of Congress Control Number: 2013936533

Springer is part of Springer Science+Business Media (www.springer.com)

To
Our Children and Grandchildren may they all live a life free of the pain and shackles of addiction:
Jeffrey, Seth, Moses, Jia, Steve, Kimberly, Kyle, Cameron, Nate, Jonah, Sarah, Eric, Jacob, Dahlia, Athena, Amaron, Jenna, Justin, Sebastian, Jesse, Giselle and Richard

Preface

The lead author of this manuscript believes that the emotional attachment of recovering addicts to 12-step programs deserves special attention. It is important that the readership of this Neuroscience Brief realize that our attempt to describe the scientific underpinnings of this powerful program will in no way reduce respect for these concepts. The concepts developed not only by "Bill W." (William G. Wilson), but by other notables such as "Dr. Bob" (Robert H. Smith) and the Oxford Group, have evolved based on empirical experiences of what worked for the Alcoholics Anonymous (AA) and later Narcotics Anonymous membership. It is with great respect and admiration that we as a group have come together in an attempt to associate each of the 12-steps with molecular, biological, and in some cases, neurogenetic explanations.

It is quite remarkable that Bill W., along with others, fought arduously against all odds to deliver the 12-step message and fellowship to alcoholics starting in Ohio, St. Louis, and New York. It is ironic that in 1939, when the Works Publishing Company, owned by a few investors through the purchase of stock, published the Big Book (*Alcoholics Anonymous*: *How Thousands of Men and Women Have Recovered from Alcoholism*), the response was less positive than expected. The end result was that Bill W. and his wife Lois lost just about everything including their house. However, realizing that the 12-step program was indeed saving lives, Bill W. never gave up!

It should be noted that Bill W. continued to smoke throughout his recovery even at 76 years of age, while dying from emphysema in the winter of 1971. From a neurological perspective, nicotine, sugar, and coffee activate and release dopamine in the *nucleus accumbens* reward site of the mesolimbic system of the brain. While not minimizing the effect AA also had on his recovery, certainly the depression Bill W. suffered during his recovery from alcohol, a 17-year battle, was indeed reduced by the continued use of all these substances. As a neuropharmacologist, I am also interested in Bill W.'s effort to combat alcoholism through biology. When he experimented with both LSD and Vitamin B3 (Niacin) therapy, many in the fellowship were dismayed with his hope to biologically assist alcoholics to gain relief from their addiction. In the mid-1980s, I was seeking support for an earlier version of a natural dopamine D2 agonist from the AA community and was similarly and emphatically turned down by the executive officers in New York City.

In an effort to expand our scientific understanding of how the 12-step program and fellowship saves lives and assists the people doing the work to better understand the role of neuroscience in addiction, we have attempted to link the remarkable benefits of each of the 12-steps with the science of molecular neurobiology and neurogenetics.

It is important to note that although the book refers to AA, due to understanding the shared neurological mechanisms between alcohol, drugs, nicotine, food and other behavioral addictions like, internet, gambling, sex and shopping, although there are some differences, the term AA has been used interchangeably with NA and any other behavioral self-help programs based on the 12 step and fellowship.

Acknowledgments

The authors would like to thank Karyn Hurley, Executive Director of the National Institute of Holistic Addiction Studies and Pamela Graham, a drug abuse counselor at Malibu Beach Recovery Center for their very knowledgeable comments and remarkable spirit. The authors are indebted to the superb editing of Margaret A. Madigan, and for the insightful and invaluable input from Ben Thompson, a Ph.D. candidate at Boston University and a 12-step philosopher. The authors would like to specially thank the following for their unconditional support: B. William Downs, Roger L. Waite, James Heaney, Mary Hauser, Thomas Simpatico, David E. Smith, Joan Borsten, and Eric R. Braverman. Marlene Oscar-Berman is the recipient of grants from the National Institutes of Health, NIAAA (RO1-AA07112 and K05-AA00219) and the Medical Research Service of the US Department of Veterans Affairs.

Contents

Chapter 1
Introduction

The purpose of this book is to provide a conceptual schematic for addiction recovery by linking the twelve-step program and fellowship with insights from neuroscience. It represents a stride toward bringing together two very different approaches to understanding a complex problem. To our knowledge, this is the first attempt to review the molecular neurobiological aspects of the twelve-step program adopted by self-help groups such as Alcoholics Anonymous (AA) and Narcotics Anonymous (NA). Of course, the hundred or so alcoholics who developed these steps by the early to late 1930s did not have the scientific tools we have today. The brain was a mysterious organ and much less was known about its workings—especially the role of neurotransmitters and reward circuitry. Through the advent of twenty-first century science and medicine, especially neuroimaging technologies, we the authors believe that science has finally caught up with both the twelve-step program and fellowship and is unraveling the amazing mysteries linked to the functioning of the brain and reward. Thus, our aim is to provide the readership with important knowledge about the many facets of neuroscience and molecular neurobiology, so that we may better understand the remarkable tool for recovery known as "The Twelve Steps." Since there is no known "cure" for addiction, it is imperative that prevention must start with the family, as suggested by Scott A. Teitelbaum in his book "Addiction: A family Affair" (2011). An understanding of the neuro-molecular biological underpinnings of the twelve steps and the work of various groups such as Al Anon as espoused in the present treatise may indeed be a new and important for continued progress toward becoming and remaining clean and sober. This information could ultimately lead to a better quality of life in recovery by incorporating principles of molecular neurobiology into working the twelve steps.

Over a half century of dedicated and rigorous scientific research into the mesolimbic system has provided insight into the addictive brain and the neurogenetic mechanisms involved in the quest for happiness. In brief, the site of the brain where one experiences feelings of well-being is called the mesocorticolimbic reward system. This is where chemical messages including serotonin (5-HT), enkephalin, gamma-aminobutyric acid (GABA), and dopamine (DA) work in concert to provide a net release of DA at the nucleus accumbens (NAc). It is well

K. Blum et al., *Molecular Neurobiology of Addiction Recovery*,
SpringerBriefs in Neuroscience, DOI: 10.1007/978-1-4614-7230-8_1,

known that genes control the synthesis, vesicular storage, metabolism, receptor formation, and catabolism of neurotransmitters (Hodge et al. 1996; Hodge and Cox 1998). The polymorphic versions of these genes have certain variations that can lead to an impairment of the neurochemical events involved in the neuronal release of DA. The cascade of these neuronal events has been termed "The Brain Reward Cascade" (Blum et al. 1990) [see Fig. 1.1a, b and 1.2]. A breakdown of this cascade will ultimately lead to the dysregulation and dysfunction of DA. Because DA has been established as the pleasure molecule and the anti-stress molecule, any reduction in function could lead to reward deficiency and result in aberrant substance-seeking behavior and a lack of wellness (Blum et al. 2000).

Humans are biologically predisposed to drink, eat, reproduce, and desire pleasurable experiences. Impairment in the mechanisms involved in reward from these natural processes lead to multiple impulsive, compulsive, and addictive behaviors governed by genetic polymorphic antecedents. Although there are a plethora of genetic variations at the level of mesolimbic activity, polymorphisms of the following candidate genes are known to predispose individuals to excessive cravings (e.g., cocaine) and can result in aberrant behaviors (Blum et al. 1996a, b, 2011b; Nestler 2005, Thomas et al. 2008). The list of genes includes the serotonergic 2A receptor (5-HTT2a); serotonergic transporter (5-HTTLPR); DRD2 receptor; DRD4 receptor; DA transporter (DAT1); the catechol-O-methyltransferase (COMT); monoamine oxidase genes. Additional predisposing influences are the genetic transcription factor DeltaFosB, CREB, the extracellular signal-regulated kinase signaling pathway, brain-derived neurotrophic factor, and glutamate transmission.

Gold and colleagues first proposed a functional DA deficit during cocaine abstinence (Dackis and Gold 1985; Gold et al. 1986). In 1996, Blum's laboratory first coined *Reward Deficiency Syndrome* (RDS) as an umbrella term for all conditions that are associated with hypodopaminergic function. Common genetic variants of the DA receptor gene (DRD2) polymorphisms (Grandy et al. 1989; Hauge et al. 1991) have been identified as putative predictors of impulsive, compulsive, and addictive behaviors (Blum et al. 1995, 1996b) [see Fig. 1.3]. For example,

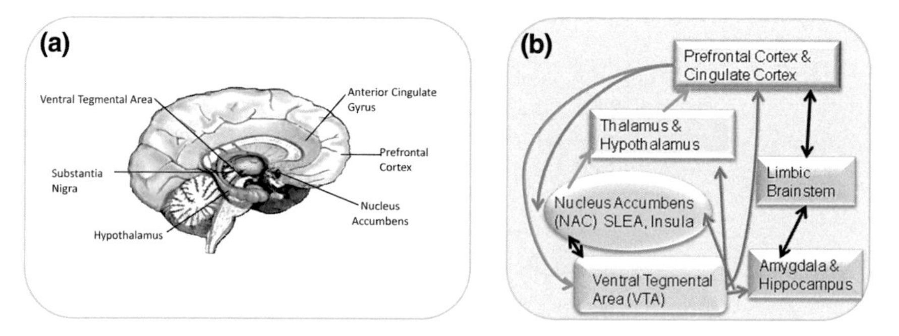

Fig. 1.1 **a** Brain reward sites. **b** Extended brain reward circuitry (Blum et al. 2012 with permission)

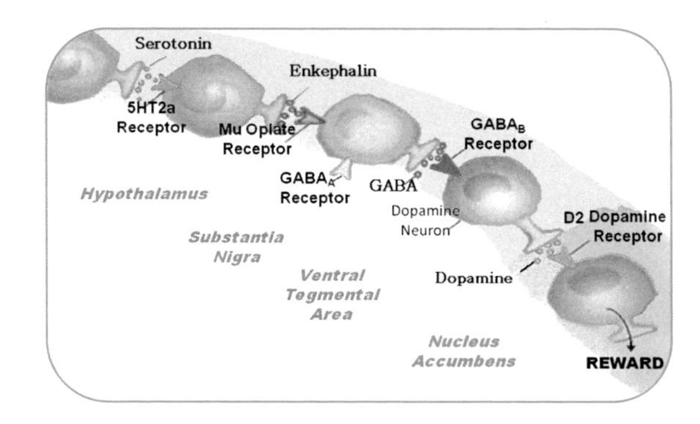

Fig. 1.2 Interaction of neurotransmitters within the mesolimbic brain reward cascade (Modified from Erickson 2007)

Fig. 1.3 Reward deficiency syndrome table

Addictive Behaviors	Impulsive Behaviors	Compulsive Behaviors	Personality Disorders
severe alcoholism	attention-deficit hyper-activity disorder (ADHD)	aberrant sexual behavior	conduct disorder
polysubstance abuse	Tourette Syndrome (TS)	Internet gaming	antisocial personality
smoking	autism	pathological gambling	aggressive behavior
obesity			

individuals who possess a paucity of serotonergic and/or dopaminergic receptors and an increased rate of synaptic DA catabolism because of high catabolic genotype of the COMT gene or high monoamine oxidase activity, are predisposed to self-medicate with any substance or behavior that will activate DA release including: alcohol, opiates, psychostimulants, nicotine, glucose, gambling, sex, and even excessive internet gaming (Comings and Blum 2000; Blum et al. 2000; Jorandby et al. 2005).

The use of most drugs of abuse, including alcohol, is associated with release of DA in the mesocorticolimbic system or reward pathway of the brain (Di Chiara and Impereto 1988) [see Fig. 1.4]. Activation of this dopaminergic system induces feelings of reward and pleasure (Dackis and Gold 1985; Volkow and Muenke 2012; Eisenberg et al. 2007). However, reduced activity of the DA system (hypodopaminergic functioning) can trigger drug-seeking behavior (Volkow 2001; Volkow et al. 2001; Gold et al. 1984; Dackis et al. 1985). Variant alleles can induce hypodopaminergic functioning through reduced DA receptor density, blunted response to DA, or enhanced DA catabolism in the reward pathway

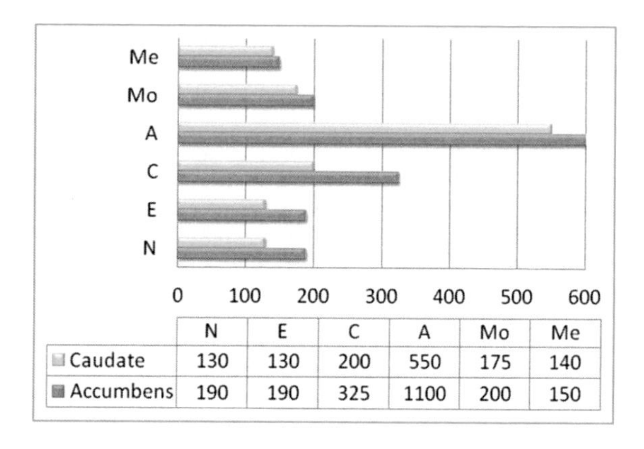

	N	E	C	A	Mo	Me
Caudate	130	130	200	550	175	140
Accumbens	190	190	325	1100	200	150

Fig. 1.4 Dopamine release. Abbreviations: *Me* (Methadone); *Mo* (Morphine); *A* (Amphetamine); *C* (Cocaine); *E* (Ethanol); *N* (Nicotine) (Di Chiara and Imperato 1988—modified)

(Hietala et al. 1994a). Cessation of chronic drug use can also induce a hypodopaminergic state that prompts drug-seeking behaviors in an attempt to address the withdrawal-induced state (Hietala et al. 1994b).

While acute use of these substances can induce a feeling of well-being, sustained and prolonged abuse unfortunately leads to a toxic "high" and results in tolerance, disease, and discomfort (Gold 1984). Thus, excessive cravings caused by carrying the DRD2 A1 allelic genotype and low DA receptors are compounded by consequential drug-seeking behavior. Conversely, normal densities of DA receptors result in low craving behaviors. In terms of preventing substance abuse or excessive glucose craving, one goal would be to induce a proliferation of DAD2 receptors in genetically prone individuals (Rothman et al. 2007). Experiments in vitro have shown that constant stimulation of the DA receptor system via a known D2 agonist in low doses results in significant proliferation of D2 receptors in spite of genetic antecedents (Boundy et al. 1995). In essence, D2 receptor stimulation signals negative feedback mechanisms in the mesolimbic system to induce mRNA expression, causing proliferation of D2 receptors. This molecular finding serves as the basis for naturally inducing DA release to cause the same induction of D2-directed mRNA and thus proliferation of D2 receptors in humans. This proliferation of D2 receptors in turn will induce the attenuation of craving behavior. In fact, this has been proven with work showing a form of gene therapy, DNA-directed overexpression of the DRD2 receptors induces a significant reduction in both alcohol and cocaine craving induced behavior in animals (Thanos et al. 2001, 2005, 2008).

These observations are the basis for the development of a functional hypothesis of drug seeking and drug use. The hypothesis is that the presence of a hypodopaminergic state, regardless of the source, is a primary cause of drug-seeking behavior. Thus, genetic polymorphisms, which induce hypodopaminergic functioning, may be the mechanism underlying a genetic predisposition to chronic drug use and relapse (Merlo and Gold 2008). Finally, utilizing the long-term dopaminergic activation approach ultimately will lead to a common, safe, and effective modality to treat RDS behaviors that include among others, substance use disorders, attention deficit–hyperactivity disorder (ADHD), and obesity.

As stated earlier, DA is known as the main neurotransmitter modulating the activation of the reward system of the brain and has been associated with pleasure (Gold 1993). While debatable, as mentioned earlier, it has been called the anti-stress molecule and the pleasure molecule (Blum et al. 1990, 2000; Comings et al. 1996; Bau et al. 2000). When DA is released into the synapse, it stimulates a number of receptors (D1–D5), which results in increased feelings of well-being and stress reduction. It is of particular interest that, the DRD2 gene has been one of the most widely studied in neuropsychiatric disorders in general and in alcoholism and other addictions in particular (Blum et al. 2000). Grasping the mechanism of motivated behavior requires an understanding of the neural circuitry of rewards (Robbins and Everitt 1996), otherwise called positive reinforcers. A positive reinforcer is operationally defined as an event known to increase the probability of a subsequent positive response, and some argue that drugs of abuse are stronger positive reinforcers than natural reinforces (e.g., food and sex) (Wightman and Robinson 2002; Epping-Jordon et al. 1998; Caine et al. 1995). The distinction between natural rewards and unnatural rewards is an important one. Natural rewards include satisfaction of physiological drives (e.g., hunger and reproduction), and unnatural rewards are learned and involve satisfaction of acquired drives (Suhara et al. 2001; James et al. 2004) for pleasure (Fort 1969) derived from alcohol and other drugs as well as from gambling and other risk-taking behaviors (Wightman and Robinson 2002; Hodge and Cox 1998; Besheer et al. 2003).

The NAc, a site within the ventral striatum, is best known for its prominent role in mediating the reinforcing effects of drugs of abuse such as cocaine (Miller and Gold 1994), alcohol (Gold 2005), nicotine (Gold and Herkov 1998), food (Blumenthal and Gold 2010), and music (Blum et al. 2010). Indeed, it is generally believed that this structure mandates motivated behaviors, such as feeding, drinking, sexual behavior, and exploratory locomotion, which are elicited by natural rewards or incentive stimuli. A basic rule of positive reinforcement is that motor responses will increase in magnitude and vigor if followed by a rewarding event. We the authors are hypothesizing a common mechanism of action for the powerful effects that drugs, music, food, and sex have on human motivation.

The human drive for the three necessary motivated behaviors, hunger, thirst, and sex, may all have common molecular genetic antecedents that, if impaired, lead to aberrant behaviors. We hypothesize that sex, like food and drugs, activates 148 brain mesolimbic reward circuitry (2009a). Moreover, dopaminergic genes and possibly other candidate neurotransmitter-related genes and their polymorphisms affect both hedonic and anhedonic behavioral outcomes. As such, we anticipate that future clinical studies involving genotyping of sex addicts will provide evidence for polymorphic associations with specific clustering of sexual typologies based on assessments using clinical instruments. We are encouraging academic and clinical scientists to embark on research coupling neuroimaging tools [i.e., functional magnetic resonance imaging (fMRI), quantitative electroencephalography (qEEG), single-photon emission computed tomography (SPECT), magnetoencephalography (MEG), and positron emission tomography (PET)] and natural dopaminergic agonistic agents (i.e., KB220Z^{TM}) to systematically target specific gene polymorphisms in the service of determining and eventually normalizing hyper- or hyposexual responses (Koob 2008, 2009; Blum et al. 2011b).

Drug microinjection studies have shown that opioids amplify the liking of sweet taste rewards. Modern neuroscience tools such as Fos plume mapping have further identified hedonic hot spots within the accumbens and pallidum, where opioids are especially tuned to magnify the liking of food rewards. Hedonic hot spots in different brain structures may interact with each other within the larger interconnected functional circuitry (Peciña et al. 2006).

Hedonic liking for sensory pleasures is an important aspect of reward, and excessive liking of particular rewards might contribute to excessive consumption and to disorders such as RDS. With this brief introduction to mesolimbic circuitry and the potential role of neurogenetics and neurotransmission for conserving a sense of well-being, we now attempt to link the twelve-step model for recovery to molecular neurobiology.

1.1 Alcoholics Anonymous/Narcotic Anonymous

AA is an international mutual aid movement that was founded in 1935 by Bill Wilson and Dr. Bob Smith (Bill W. and Dr. Bob) in Akron, Ohio. AA's name derived from its first book, informally called "The Big Book," originally titled *AA: The Story of How More than One-hundred Men Have Recovered from Alcoholism*. In the Big Book, the twelve steps are listed in a chapter entitled "How It Works." The "it" in the title refers to the program of the twelve steps to recovery, and the objective of this endeavor is primarily descriptive, not explanatory. In other words, the book describes how the first AA members obtained a spiritual experience, rather than explaining how the program works by recourse to the attendant mechanisms of change—which is germane to the objectives of the present treatise. In any case, although the word "recovery" may not be a realistic term, certainly based on our knowledge today, it does reflect success in restraint from alcohol *per se*, but not from all addictions.

Wilson and Smith, with other early members, developed AA's twelve-step program of spiritual and character development. The AA's preamble, read at the beginning of many meetings, states: "Our primary purpose is to stay sober and help other alcoholics achieve sobriety." It is noteworthy that during this period, Margaret Mead discussed cultural patterns in a global sense and began to lay down a framework for character as it relates to medicine and even schizophrenia (Mead 1953). AA's twelve traditions were introduced in 1946 to help AA strengthen and grow. The traditions recommend that members and groups remain anonymous in public media, altruistically help other alcoholics, and include all who wish to stop drinking. The traditions also recommend that AA members acting on behalf of the fellowship steer clear of dogma, governing hierarchies, and involvement in public issues. Subsequent fellowships such as NA have adopted and adapted the twelve steps and the twelve traditions to their respective primary purposes (Chappel and Dupont 1999).

Since 1935, AA has spread "across diverse cultures holding different beliefs and values," including geopolitical areas resistant to grassroot movements. AA estimates it has more than two million members. While there is a difference between the twelve-step program and AA/NA fellowship, both are important for successful recovery. Here, we interchange the words "fellowship" and "program" because there are those who believe they are synonymous.

AA generally avoids discussing the medical nature of alcoholism nonetheless AA is regarded as the basis for "What is recovery?" (Betty Ford Institute Consensus Panel 2007) and "What is the disease theory of alcoholism (Blum 1991)?" The American Psychiatric Association has recommended sustained treatment in conjunction with AA's programs, or similar community resources, for chronic alcoholics unresponsive to brief treatment. The twelve steps are the essential component for the high recovery rates for physicians (DuPont et al. 2009), and other professionals, as well as non-professional individuals (Miller and Gold 1991). It is noteworthy, however, that data indicate up to a 64 % drop-out rate in the first year (Timko and Debenedetti 2007). This estimate is approximate, since AA does not keep formal records of members in the fellowship as a whole. Moreover, individual AA groups do not keep record of attrition rates, because alcoholics move or change groups, etc.

It is important to keep in mind that AA is a complex social system that may be understood in two essential respects: (1) a formal organized hierarchy, represented by official literature, and (2) an organic network of individuals mutually identifying with one another's experiences. The significance of this distinction is that there is nothing such as "real AA." The fellowship is a self-selecting population that diversifies in virtually every sociocultural direction, according to the niche available to be filled. This is increasingly true of the growth and spread of other twelve-step movements. It would be inaccurate to declare a single twelve-step position on anything, since each member within each group within each fellowship will likely have modified the program in some way or another to suit his/her or its attendant needs. Even official literature may not be representative of the views of a group or fellowship's individual members. Ultimately, these are issues that probably best lend themselves to sociological methods of inquiry.

References

Bau CH, Almeida S, Hutz MH (2000) The TaqI A1 allele of the dopamine D2 receptor gene and alcoholism in Brazil: association and interaction with stress and harm avoidance on severity prediction. Am J Med Genet 12; 96(3):302–306

Besheer J, Cox AA, Hodge CW (2003) Coregulation of ethanol discrimination by the nucleus accumbens and amygdala. Alcohol Clin Exp Res 27(3):450–456

Betty Ford Institute Consensus Panel (2007) What is recovery? A working definition from the Betty Ford Institute. J Subst Abuse Treat 33:221–228

Blum K (with Payne JE) (1991) Alcohol and the addictive brain: new hope for alcoholics from biogenetic research. The Free Press Simon and Schuster, Inc, New York ISBN 0-02-903701-8

Blum K, Noble EP, Sheridan PJ, Montgomery A, Ritchie T, Jagadeeswaran P, Nogami H, Briggs AT, Cohn JB (1990) Allelic association of human dopamine D2 receptor gene in alcoholism. JAMA 263:2055–2060

Blum K, Wood RC, Braverman ER, Chen TJ, Sheridan PJ (1995) The D2 dopamine receptor gene as a predictor of compulsive disease: Bayes' theorem. Funct Neurol 10(1):37–44

Blum K, Braverman ER, Wood RC, Gill J, Li C, Chen TJ, Taub M, Montgomery AR, Sheridan PJ, Cull JG (1996a) Increased prevalence of the Taq I A1 allele of the dopamine receptor gene (DRD2) in obesity with comorbid substance use disorder: a preliminary report. Pharmacogenetics 6(4):297–305

Blum K, Sheridan PJ, Wood RC, Braverman ER, Chen TJ, Cull JG, Comings DE (1996b) The D2 dopamine receptor gene as a determinant of reward deficiency syndrome. J R Soc Med 89(7):396–400

Blum K, Braverman ER, Holder JM, Lubar JF, Monastra VJ, Miller D, Lubar JO, Chen TJ, Comings DE (2000) Reward deficiency syndrome: a biogenetic model for the diagnosis and treatment of impulsive, addictive, and compulsive behaviors. J Psychoactive Drugs 32(i–iv):1–112

Blum K, Chen TJH, Chen ALH, Madigan M, Downs BW, Waite RL, Braverman ER, Kerner M, Bowirrat A, Giordano J, Henshaw H, Gold MS (2010) Do dopaminergic gene polymorphisms affect mesolimbic reward activation of music listening response? Therapeutic impact on reward deficiency syndrome (RDS). Med Hypotheses 74(3):513–520

Blum K, Miller M, Perrine K, Liu Y, Giordano J, Oscar-Berman M (2011b) Mesolimbic hypodopaminergic function a potential nutrigenomic therapeutic target for drug craving and relapse. Abstract, September 2011, XIX World Congress of Psychiatric Genetics, Washington, DC, p 248

Blum K, Oscar-Berman M, Stuller E, Miller D, Giordano J, Morse S, McCormick L, Downs WB, Waite RL, Barh D, Neal D, Braverman ER, Lohmann R, Borsten J, Hauser M, Han D, Liu Y, Helman M, Simpatico T (2012) Neurogenetics and nutrigenomics of neuro-nutrient therapy for reward deficiency syndrome (RDS): clinical ramifications as a function of molecular neurobiological mechanisms. J Addict Res Ther 3:139. doi:10.4172/2155-6105.1000139

Blumenthal DM, Gold MS (2010) Neurobiology of food addiction. Curr Opin Clin Nutr Metabolic Care 13:359–365

Boundy VA, Pacheco MA, Guan W, Molinoff PB (1995) Agonists and antagonists differentially regulate the high affinity state of the D2L receptor in human embryonic kidney 293 cells. Mol Pharmacol 48(5):956–964)

Caine SB, Heinrichs SC, Coffin VL, Koob GF (1995) Effects of the dopamine D-1 antagonist SCH 23390 microinjected into the accumbens, amygdala or striatum on cocaine self-administration in the rat. Brain Res 18; 692(1–2):47–56

Chappel JN, Dupont RL (1999) Twelve-step and mutual-help programs for addictive disorders. psychiatric clinics of North America 22 (2):425–446. PMID 10385942. doi:10.1016/S0193-953X(05)70085-X

Comings DE, Blum K (2000) Reward deficiency syndrome: genetic aspects of behavioral disorders. Prog Brain Res 126:325–341

Comings DE, Ferry L, Bradshaw-Robinson S, Burchette R, Chiu C, Muhleman D (1996) The dopamine D2 receptor (DRD2) gene: a genetic risk factor in smoking. Pharmacogenetics 6(1):73–79

Dackis CA, Gold MS (1985) New concepts in cocaine addiction: the dopamine depletion hypothesis. Neurosci Biobehav Rev 9(3):469–477

Dackis CA, Gold MS, Davies RK, Sweeney DR (1985) Bromocriptine treatment for cocaine abuse: the dopamine depletion hypothesis. Int J Psychiatry Med 15(2):125–135

Di Chiara G, Imperato A (1988) Drugs abused by humans preferentially increase synaptic dopamine concentrations in the mesolimbic system of freely moving rats. Proc Natl Acad Sci USA 85(14):5274–5278

DuPont RL, McLellan AT, White WL, Merlo LJ, Gold MS (2009) Setting the standard for recovery: physicians' health programs. J Subst Abuse Treat 36(2):159–171

Eisenberg DT, Campbell B, Mackillop J, Lum JK, Wilson DS (2007) Season of birth and dopamine receptor gene associations with impulsivity, sensation seeking and reproductive behaviors. PLoS One 21, 2(11):e1216

Epping-Jordan MP, Markou A, Koob GF (1998) The dopamine D-1 receptor antagonist SCH 23390 injected into the dorsolateral bed nucleus of the stria terminalis decreased cocaine reinforcement in the rat. Brain Res 16; 784(1–2):105–115

Erickson (2007) The science of addiction. W.W. Norton & Co, New York

Fort J (1969) The pleasure seekers: the drug crisis, youth and society. Bobbs-Merrill, Indianapolis

Gold MS (1984) 800 Cocaine. Bantam Books, New York, NY

Gold MS (2005) Alcohol abuse. Psychiatric Annals 35(6):458–460

Gold MS (1993) Opiate addiction and the locus coeruleus. The clinical utility of clonidine, naltrexone, methadone, and buprenorphine. Psychiatr Clin North Am. 16(1):61–73

Gold MS, Herkov MJ (1998) Tobacco smoking and nicotine dependence: biological basis for pharmacotherapy from nicotine to treatments that prevent relapse. J Addict Dis 17(1):7–22

Gold MS, Dackis CA, Washton AM (1984) The sequential use of clonidine and naltrexone in the treatment of opiate addicts. Adv Alcohol Subst Abuse 3(3):19–39

Gold MS, Dackis CA, Pottash ALC, Extein I, Washton A (1986) Cocaine update: from bench to bedside. In: Stimmel B (ed) Advances in alcohol and substance abuse. Haworth Press, Binghamton, NY, pp 35–60

Grandy DK, Litt M, Allen L, Bunzow JR, Marchionni M, Makam H, Reed L, Magenis RE, Civelli O (1989) The human dopamine D2 receptor gene is located on chromosome 11 at q22–q23 and identifies a TaqI RFLP. Am J Hum Genet 45(5):778–785

Hauge XY, Grandy DK, Eubanks JH, Evans GA, Civelli O, Litt M (1991) Detection and characterization of additional DNA polymorphisms in the dopamine D2 receptor gene. Genomics 10(3):527–530

Hietala J, Syvälahti E, Vuorio K, Någren K, Lehikoinen P, Ruotsalainen U, Räkköläinen V, Lehtinen V, Wegelius U (1994a) Striatal D2 dopamine receptor characteristics in neuroleptic-naive schizophrenic patients studied with positron emission tomography. Arch Gen Psychiatry 51(2):116–123

Hietala J, West C, Syvälahti E, Någren K, Lehikoinen P, Sonninen P, Ruotsalainen U (1994b) Striatal D2 dopamine receptor binding characteristics in vivo in patients with alcohol dependence. Psychopharmacology 116(3):285–290

Hodge CW, Cox AA (1998) The discriminative stimulus effects of ethanol are mediated by NMDA and GABA(A) receptors in specific limbic brain regions. Psychopharmacology 139(1–2):95–107

Hodge CW, Chappelle AM, Samson HH (1996) Dopamine receptors in the medial prefrontal cortex influence ethanol and sucrose-reinforced responding. Alcohol Clin Exp Res 20(9):1631–1638

James GA, Gold MS, Liu Y (2004) Interaction of satiety and reward response to food stimulation. J Addict Dis 23(3):23–37

Jorandby L, Frost-Pineda K, Gold MS (2005) Addiction to food and brain reward systems. Sexual Addict Compulsivity 12(2–3):201–217

Koob GF (2008) Hedonic homeostatic dysregulation as a driver of drug-seeking behavior. Drug Discov Today Dis Models 5(4):207–215

Koob GF (2009) Neurobiological substrates for the dark side of compulsivity in addiction. Neuropharmacology 56(Suppl 1):18–31

Mead M (1953) Cultural patterns and technical change. UNESCO, Paris

Merlo LJ, Gold MS (2008) Prescription opioid abuse and dependence among physicians: hypotheses and treatment. Harv Rev Psychiatry 16(3):181–194

Miller NS, Gold MS (1991) Drugs of abuse: A comprehensive series for clinicians, vol II. Alcohol Plenum Medical Book Company, New York and London

Miller NS, Gold MS (1994) Dissociation of "conscious desire" (craving) from and relapse in alcohol and cocaine dependence. Ann Clin Psychiatry 2:99–106

Nestler EJ (2005) The neurobiology of cocaine addiction. Sci Pract Perspect 3(1):4–10

Peciña S, Smith KS, Berridge KC (2006) Hedonic hot spots in the brain. Neuroscientist 12(6):500–511
Robbins TW, Everitt BJ (1996) Neurobehavioural mechanisms of reward and motivation. Curr Opin Neurobiol 6(2):228–236
Rothman RB, Blough BE, Baumann MH (2007) Dual dopamine/serotonin releasers as potential medications for stimulant and alcohol addictions. AAPS J 5;9(1):E1–E10
Suhara T, Yasuno F, Sudo Y, Yamamoto M, Inoue M, Okubo Y, Suzuki K (2001) Dopamine D2 receptors in the insular cortex and the personality trait of novelty seeking. Neuroimage 13(5):891–895
Teitelbaum S (2011) Addiction is a family affair. University Press of Florida, Gainesville
Thanos PK, Volkow ND, Freimuth P, Umegaki H, Ikari H, Roth G, Ingram DK, Hitzemann R (2001) Overexpression of dopamine D2 receptors reduces alcohol self-administration. J Neurochem 78(5):1094–1103
Thanos PK, Rivera SN, Weaver K, Grandy DK, Rubinstein M, Umegaki H, Wang GJ, Hitzemann R, Volkow ND (2005) Dopamine D2R DNA transfer in dopamine D2 receptor-deficient mice: effects on ethanol drinking. Life Sci 27; 77(2):130–139
Thanos PK, Michaelides M, Umegaki H, Volkow ND (2008) D2R DNA transfer into the nucleus accumbens attenuates cocaine self-administration in rats. Synapse 62(7):481–486
Thomas MJ, Kalivas PW, Shaham Y (2008) Neuroplasticity in the mesolimbic dopamine system and cocaine addiction. Br J Pharmacol 154(2):327–342
Timko C, Debenedetti A (2007) A randomized controlled trial of intensive referral to 12-step self-help groups: one-year outcomes. Drug Alcohol Dependence 90(2–3):270–279, PMID 17524574. doi:10.1016/j.drugalcdep.2007.04.007
Volkow ND (2001) Drug abuse and mental illness: progress in understanding comorbidity. Am J Psychiatry 158(8):1181–1183
Volkow ND, Chang L, Wang GJ, Fowler JS, Ding YS, Sedler M, Logan J, Franceschi D, Gatley J, Hitzemann R, Gifford A, Wong C, Pappas N (2001) Low level of brain dopamine D2 receptors in methamphetamine abusers: association with metabolism in the orbitofrontal cortex. Am J Psychiatry 158(12):2015–2021
Volkow ND, Muenke M (2012) The genetics of addiction. Hum Genet. 131(6):773–777. doi: 10.1007/s00439-012-1173-3
Wightman RM, Robinson DL (2002) Transient changes in mesolimbic dopamine and their association with 'reward'. J Neurochem 82(4):721–735

Chapter 2
Molecular Neurobiology of Recovery with the Twelve Steps

2.1 Step 1: We Admitted that We were Powerless Over Alcohol—that Our Lives had Become Unmanageable

At the outset, it may be helpful to point out that to fully understand how AA works, it must be taken in, and on its own terms. Moreover, AA's ideology of "alcoholism" is subtly different from the clinical interpretation of "addiction." Understanding this distinction, and especially AA's interpretation, is crucial to the task of unpacking its twelve steps. Indeed, there is no official definition of alcoholism in AA literature. Rather, it is thought of as an illness that only a spiritual experience can conquer. The "powerlessness" that "*we* admitted" is a personal matter that cannot be reduced to the nature of "addiction" *per se*, but rather, to the nature of the "alcoholic"—hence, "*we* admitted." This description is akin to William James' (1902) description of the "sick souled" individual in *The Varieties of Religious Experiences*. This is significant because it leaves one's status as an alcoholic up to the interpretation of one's own experience with alcohol. In other words, it is a process of identifying with the description of being *alcoholic*, rather than the definition of *alcoholism*.

According to self-help groups, in order to reach sobriety, individuals must accept the fact that they are powerless over their addiction. As such, dominant addiction interventions conceptualize the addict as powerless, due to either moral or physical weakness (Gowan et al. 2012). While the premise may be true, others in the field have rejected this concept, suggesting that it is a defeatist attitude that will backfire in recovery. Instead, they suggest "readiness" along with a positive attitude. Accordingly, clinicians have reported that their clients frequently misinterpret an inferred "disease model" as meaning that they are powerless victims of addiction. Clients' readiness to change can be enhanced by providing them with a coherent method of conceptualizing the task of rehabilitation, one that explains how change is possible in spite of the disease model of addiction. No one person is doomed and hopeless. This is preferable to a possibly confusing mix of self-help, "bootstrap" philosophy, and "disease" talk available in many rehabilitation programs. Productive methods of dealing with clients' guilt seem parsimonious, along

K. Blum et al., *Molecular Neurobiology of Addiction Recovery*,
SpringerBriefs in Neuroscience, DOI: 10.1007/978-1-4614-7230-8_2,

with the need to develop a positive purpose in life using positive psychological principles, as an essential aspect of effective relapse management (Ford 1996).

2.1.1 Loss of Control (Powerlessness) and Molecular Neurobiology

Addiction is a chronic relapsing disorder associated with compulsive drug taking, drug seeking, and a loss of control in limiting intake (Greene and Gold 2012). Koob and Volkow (2010) conceptualized drug addiction as a disorder that involves elements of both impulsivity and compulsivity that together yield a composite three-stage cycle: binge/intoxication, withdrawal/negative affect, and preoccupation/anticipation (craving). Neuroimaging studies of human and non-human animals have revealed discrete circuits that mediate the three stages of the addiction cycle (Goldstein and Volkow 2011). Key elements are the ventral tegmental area and ventral striatum, a focal point for the binge/intoxication stage. The extended amygdala has a key role for the withdrawal/negative affect stage. Disrupted inhibitory control and the preoccupation/anticipation stage are controlled by a widely distributed network involving the orbitofrontal cortex–dorsal striatum, dorsolateral prefrontal and inferior frontal cortices, cingulate gyrus, basolateral amygdala, hippocampus, and insula (Koob and Volkow 2010).

Addiction also has been conceptualized as a shift from controlled experimentation to uncontrolled, compulsive patterns of use (Herkov et al. 2013). Current neurobiological models of addiction emphasize changes within the brain's reward system, such that drugs of abuse "hijack" this system and bias behavior toward further drug use.

A review of the current literature suggests that the loss of newly born progenitor cells, particularly in hippocampal and cortical brain regions, plays a role in determining vulnerability to relapse in rodent models of drug addiction. Most recently, Mandyam and Koob (2012) found that "normalization" of drug-impaired neurogenesis or gliogenesis may help reverse pathological neuroplasticity during abstinence and thus may help reduce the vulnerability to relapse and enhance recovery.

While this model explains the involuntary nature of craving and the motivational drive to continue drug use, it does not fully explain why addicted individuals are unable to control their drug use when faced with potentially disastrous consequences. It has been argued by Hyman (2007) that such maladaptive and uncontrolled behavior is underpinned by a failure of the brain's inhibitory control mechanisms, which are complex. Hyman goes on to refer to addiction as a "disease of learning and memory." This is potentially relevant to AA's own approach, since a good deal of attention is given to what they call the "strange mental blank spots" they encounter in the face of temptation to drink. For this reason, one of AA's most celebrated practices is that of storytelling about one's experiences as an alcoholic. Storytelling in this format functions as a convenient mnemonic and helps to constantly upgrade memory, in order to fend off those mental blank spots on future occasions. The current literature suggests the brain's reward system, which includes temporal lobe memory centers, and prefrontal brain circuitry critical in inhibitory control over behavior are dysfunctional in addicted individuals (Goldstein and Volkow 2011; Harris et al. 2008; Makris et al. 2008). These same regions have been implicated in other compulsive conditions characterized by deficits in inhibitory control over maladaptive behaviors, such as obsessive–compulsive disorder (Kienast et al. 2008; Yaryura-Tobias et al. 1995).

Moreover, Koob and Le Moal (2008) hypothesized that the key neurochemical elements involved in reward and stress within basal forebrain structures (the ventral striatum and extended amygdala) are dysregulated in addiction, so that they convey the opponent motivational processes that drive dependence. Specific neurochemical elements in these structures include not only decreases in reward neurotransmission such as DA and opioid peptides in the ventral striatum, but also recruitment of brain stress systems such as corticotrophin-releasing factor (CRF), noradrenaline, and dynorphin in the extended amygdala. Acute withdrawal from all major drugs of abuse produces increases in reward thresholds, anxiety-like responses, and extracellular levels of CRF in the central nucleus of the amygdala. CRF receptor antagonists block excessive drug intake produced by dependence. A brain stress response system is hypothesized to be activated by acute excessive drug intake, to be sensitized during repeated withdrawal, to persist into protracted abstinence, and to contribute to stress-induced relapse. The combination of loss of reward function and recruitment of brain stress systems provides a powerful neurochemical basis for the long hypothesized opponent motivational processes responsible for the negative reinforcement driving addiction (Koob and Le Moal 2008; Broderick et al. 1973).

We have proposed (Blum and Gold 2011) that in chronically addicted individuals, maladaptive behaviors and high relapse rates may be better conceptualized as being compulsive in nature as a result of dysfunction within inhibitory brain circuitry, particularly during symptomatic states. Specifically, we are proposing that recent studies have indicated that genetic, personality, and environmental factors are predictors of drug use in adolescents. Exploration of various treatment approaches for the most part has revealed poor outcomes in terms of relapse prevention and continued drug hunger (Blum and Gold 2011). The core reward circuitry consists of an in-series circuit linking the ventral tegmental area, NAc, and ventral pallidum via the medial forebrain bundle. Although originally believed just to encode the set point of hedonic tone, these circuits now are believed to be functionally far more complex, also encoding attention, expectancy of reward, disconfirmation of reward expectancy, and incentive motivation. This model may help to explain why some addicts lose control over their drug use and engage in repetitive self-destructive patterns of drug seeking and drug taking that take place at the expense of other important activities. This model may also have clinical utility, as it allows for the adoption of treatments effective in other disorders of inhibitory dysregulation an area of active research.

According to Gardner (2011), hedonic dysregulation within these circuits may lead to addiction. He suggests that the "second-stage" dopaminergic component in this reward circuitry is the crucial addictive drug–sensitive component. This is so because all addictive drugs have in common that they enhance (directly, indirectly, or even trans-synaptically) dopaminergic reward function in the NAc. Drug self-administration is partially regulated by NAc DA levels and is done to keep NAc DA within a specific elevated range to maintain a desired hedonic level. For some classes of addictive drugs (e.g., opiates), tolerance to the euphoric effects develops with chronic use. Post-use dysphoria then comes to dominate reward circuit hedonic tone and addicts no longer use drugs to get high, but to get back to normal (get straight). Accordingly, the brain circuits mediating the pleasurable effects of addictive drugs are anatomically, neurophysiologically, and neurochemically different from those mediating physical dependence and different from those mediating craving and relapse.

Hedonic levels also might be conceived externally, as well as internally. Hedonism, by its nature, is about the pursuit of something not present. A hedonic treadmill (Brickman and Campbell 1971) is as much about attaining external goals as internal goals. Of course, it is a "treadmill," because there is always something yet to be attained, which can potentially result in the experience of despair, disappointment, or frustration. This is relevant to AA, because there are scores of personal stories told by alcoholics within the fellowship who describe their primary problem in precisely these terms. It certainly describes the second part of the first step "...that our lives had become unmanageable."

There are important genetic variations in vulnerability to drug addiction, yet environmental factors such as stress and social defeat also alter brain reward mechanisms in such a manner as to impart vulnerability to addiction. In short, a bio-psycho-social model of etiology holds very well for addiction. Moreover, addiction appears to correlate with a hypodopaminergic dysfunctional state within the reward circuitry of the brain. Neuroimaging studies in humans add credence to this hypothesis. Credible evidence also implicates serotonergic, opioid, endocannabinoid, GABAergic, and glutamatergic mechanisms as described in the well-known *Brain Reward Cascade* in addiction (Blum and Kozlowski 1990) (see Fig. 1.2).

Drug addiction progresses from occasional recreational use to impulsive use to habitual compulsive use. That is, liking the drug moves on to wanting the drug, which then moves on to needing the drug. This progression correlates with a progression from reward-driven to habit-driven drug-seeking behaviors. The three classical sets of craving and relapse triggers are (1) re-exposure to addictive drugs, (2) stress, and (3) re-exposure to environmental cues (people, places, things) previously associated with drug-taking behavior. This behavioral progression correlates with a neuroanatomical progression from ventral striatal (NAc) to dorsal striatal control over drug-seeking behavior. That is, drug-triggered relapse involves the NAc and the neurotransmitter DA (Blum et al. 2011b; Miller and Gold 1994). Stress-triggered relapse involves the central nucleus of the amygdala, the bed nucleus of the stria terminalis, and the neurotransmitter corticotrophin-releasing factor, as well as the lateral tegmental noradrenergic nuclei of the brain stem and the neurotransmitter norepinephrine. Cue-triggered relapse involves the basolateral nucleus of the amygdala, the hippocampus, and the neurotransmitter glutamate.

At this point, it is important to clarify AA's solution to the third trigger of craving and relapse. In the *Big Book*, it states, "Assuming we are spiritually fit, we can do all sorts of things alcoholics are not supposed to do. People have said we must not go where liquor is served; we must not have it in our homes; we must shun friends who drink; we must avoid moving pictures which show drinking scenes; we must not go into bars; our friends must hide their bottles if we go to their houses; we mustn't think or be reminded about alcohol at all. Our experience shows that this is not necessarily so. We meet these conditions every day. An alcoholic, who cannot meet them, still has an alcoholic mind; there is something the matter with his spiritual status. In our belief any scheme of combating alcoholism which proposes to shield the sick man from temptation is doomed to failure."

Nevertheless, knowledge of the neuroanatomy, neurophysiology, neurochemistry, and neuropharmacology of addictive drug action in the brain is currently producing a variety of strategies for pharmacotherapeutic and genomic treatment of drug addiction when needed, some of which appear promising. Blum and Gold (2011) are proposing a paradigm shift involving the incorporation of genetic testing in residential, non-residential, and aftercare environments, to identify risk alleles coupled with D2 receptor stimulation using neuroadaptogen amino acid precursor enkephalinase–COMT inhibition therapy. A natural but therapeutic nutraceutical formulation that potentially induces DA release could cause the induction of D2-directed mRNA and proliferation of D2 receptors. We further hypothesize that this proliferation of D2 receptors in turn will induce the attenuation of drug-like craving behavior. The success of pharmacological therapies in contrast has been limited because these powerful agents have focused on maintenance or interference with drug euphoria rather than correcting or compensating for pre-morbid DA system deficits. These concepts await further confirmation via required neuroimaging studies.

Other tactics may include transcranial magnetic stimulation (Diana 2011). Accordingly, morphological evaluations fed into realistic computational analysis of the medium spiny neuron of the NAc, postsynaptic counterpart of DA terminals, show profound changes in structure and function of the entire mesolimbic system. In line with these findings, human neuroimaging studies have shown a reduction in DA receptors accompanied by a lesser release of endogenous DA in the ventral striatum of cocaine, heroin, and alcohol-dependent subjects, thereby offering visual proof of the DA-impoverished addicted human brain. The lasting reduction in physiological activity of the DA system led to the idea that an increment in its activity, to restore pre-drug levels, may yield significant clinical improvements (reduction in craving, relapse, and drug seeking/taking). In theory, it may be achieved pharmacologically and/or with novel interventions such as transcranial magnetic stimulation.

While the concept of powerlessness may be controversial in the field, the first step admitting personal powerlessness over addiction is supported by the actual mechanisms involved in the neurobiological circuits of our brain. It begins with genetic vulnerability to addiction and is compounded by epigenetically induced environmental elements. Stress and the toxic effects of the drugs and compulsive behaviors themselves induce changes in the neuroanatomy, neurophysiology, and neurochemistry of the brain that effect hedonic tone, physical dependence, craving, and relapse. In essence, it is very true that indeed a person is powerless and has no control over drug-seeking and other damaging behaviors in spite of denial of their loss of control over drug abuse and erroneous thoughts concerning their "pseudo-power" over their unwanted behavior.

2.1.2 Unmanageable

The evolutionary significance of neurochemical events in the brain has received minimal attention in the field of addiction research. Likewise, a dearth of neuro-scientific research postulating how basic brain circuits might mediate emotional urges has retarded the development of scientific perspectives that could inform new inquiries into the underlying dynamics and treatment of addictions (Panksepp et al. 2002).

Admitting that one is powerless over addiction is in itself an emotional response. Emotional responses are linked to emotional urges that activate brain systems and result in an inability to manage (control) drug-seeking urges. In this regard, there are at least two emotional systems involved: reward seeking and separation distress.

Specifically, drug addiction may be regarded as the disease of the brain reward system [American Society of Addiction Medicine] (Smith 2012). This system, closely related to the system of emotional arousal, is located predominantly in the limbic structures of the brain. The existence and location of the "pleasure centers," were established originally by a demonstration of limbic structures as the location from which electrical self-stimulation can be readily evoked in non-human animals. The main neurotransmitter involved in the reward is DA, but other monoamines and acetylcholine may also participate. The dopaminergic neurons of the ventral tegmental area that project to the NAc, amygdala, prefrontal cortex, and other forebrain structures are the anatomical core of the reward system.

Moreover, Vetulani (2001) suggested that the discovery of cocaine- and amphetamine-related transcript peptides may importantly expand our knowledge about the neurochemistry of reward. Natural rewarding activities and artificial chemical rewarding stimuli act at the same locations, but while natural activities are controlled by feedback mechanisms that activate aversive centers, no such restrictions bind the responses to artificial stimuli. Certainly, the powerful effects of drugs and alcohol on brain reward pathways and associated neurotransmitter levels (synthesis, storage, and catabolism) are well known. For example, long-term abuse of alcohol has been shown to significantly reduce brain endorphins (Blum et al. 1982a) that will ultimately influence brain reward by losing control of GABA neurotransmission and as such reduce DA release in the NAc inducing profound unmanageable craving behavior. Specifically, Golden Syrian hamsters were placed individually in cages with three drinking bottles: one empty, containing water, and the third containing water and ethanol. Control hamsters received water only. After one year, the experimental hamsters showed a significantly lower concentration of leucine-enkephalin-like immunoreactive substance in the basal ganglia than the control hamsters. This finding showed that the action of ethanol involves endogenous peptidyl opiates (Blum et al. 1982a).

Interestingly, acute ethanol administration (Fig. 2.1) increased methionine-enkephalin (met-enkephalin) and beta-endorphin levels in distinct areas of the rat brain, whereas chronically supplied ethanol caused a depression of met-enkephalin and beta-endorphin levels in most of the brain areas investigated (Blum et al. 1982b). The beta-endorphin content of the intermediate/posterior lobe of the pituitary of rats and guinea pigs was decreased by 70 %. Withdrawal of ethanol resulted in a complete recovery of endorphin levels in brain and pituitary within 2 weeks (Schulz et al. 1980). Moreover, similar findings were reported by Wüster et al. (1980) for the effect of morphine on endorphin synthesis. Turchan et al. (1997) also found that repeated morphine administration to mice resulted in a reduced mRNA expression of PENK (enkephalin precursor) in the NAc only and not in the striatum. This effect was also reported by Turchan et al. (1997) for D2 mRNA levels in the NAc only. Interestingly, the acute treatment of rats with diazepam induced pronounced changes in brain enkephalin concentrations as was estimated for methionine (met)-enkephalin and in some representative experiments for leucine (leu)-enkephalin, employing highly specific radioimmunoassays (Fig. 2.1).

Diazepam selectively increased the enkephalin concentrations in the hypothalamus by about 35 % and lowered it in the corpus striatum by roughly 25 %; no changes could be detected in the medulla oblongata/pons or midbrain (Duka et al. 1979). The authors suggest that an increase in enkephalin concentrations in the hypothalamus may be important to understand the anti-stress effects of Benzodiazepines.

Chronic cocaine administration to animals also resulted in a significant down-regulation of mu-opiate receptors in limbic cortical layer 3 (17 % lower than saline-treated controls, $p = 0.03$), the core of the NAc (16 % decrease, $p = 0.05$), and the nucleus of the diagonal band (18 % decrease, $p = 0.05$). The mu receptor may manifest, as do other neural markers (e.g., DA transporter, DA efflux), a biphasic temporal pattern with up-regulation during early phases of cocaine withdrawal but a down-regulation at later times (Sharpe et al. 2000).

In terms of an unmanageable life due to addiction to psychoactive substances like alcohol, morphine, diazepam, and cocaine, these substances activate powerfully the reward system and they may produce addiction, which in humans is a chronic, recurrent disease, characterized by absolute dominance of drug-seeking behavior. The craving induced by substances of addiction inhibit other behaviors, including the ability to accept God and/or become spiritual. The adaptation of an organism to a chronic intake of drugs involves development of adaptive changes, sensitization, or tolerance. It is thought that the gap between sensitization developing for the incentive value of the drug and tolerance to the reward induced by its consumption underlies

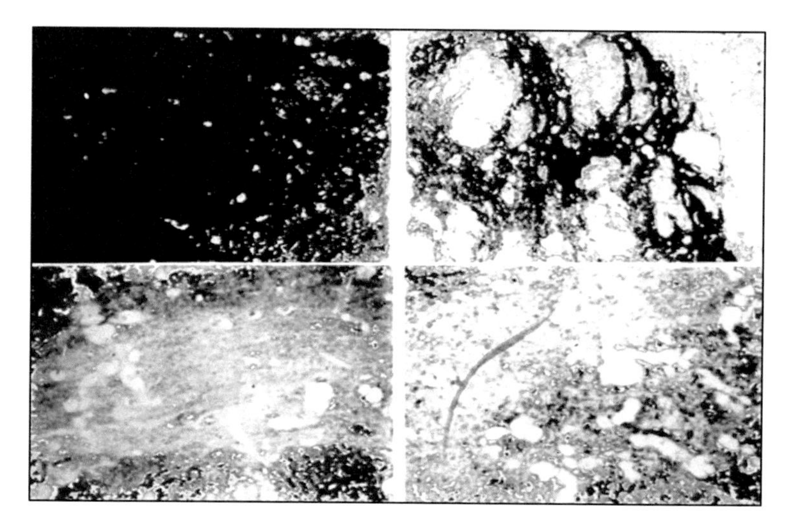

Fig. 2.1 Densities of leucine-enkephalin immune-reactive substances in basal ganglia of hamster brains. Brain sections showing relative densities of leucine-enkephalin immune-reactive substances in the basal ganglia of hamsters that consumed ethanol (*left*) and controls that did not (*right*). Each picture represents a different animal (from Blum et al. Science 1982b, with permission)

the vicious circle of events leading to drug dependence. At this point, it should be noted that according to AA literature, the only working definition of an alcoholic provided by AA literature is someone who cannot stop drinking *and stay stopped.* The evidence for one's powerlessness over alcohol is the physiological craving one experiences upon consumption of the substance, as well as his/her subsequent relapses, which could occur days, months, or years after his/her last drink *and may be due to genetically induced reward circuitry impairment.*

Although genetic factors play a very significant role in the process of addiction and especially in risk for developing reward dependence behaviors, as we see from experiments published above, there are strong epigenetic effects of powerful substances. Those substances profoundly affect brain reward homeostasis and an unmanageable desire to self-administer drugs of abuse. This ultimately leads to powerlessness, an inability to control behaviors that influences every aspect of one's life.

2.2 Step 2: Came to Believe that a Power Greater than Ourselves Could Restore Us to Sanity

Since the discovery of the double helix, the study of brain function, in terms of both physiology and behavioral traits, has resulted in a plethora of research linking these activities to the genetic basis of neurotransmitter function. Knowledge about how genes are expressed, as well as their potential impairment due to polygenic inheritance, can shed light on predispositions to addiction and self-destructive behaviors. Information derived from scientific explorations of genetic traits may have important links to understanding the basis for feelings of well-being and potentially the phenomena associated with human happiness. While non-genetic-oriented research of social, political, and biological studies has addressed the impact of social and institutional environments on mass political attitudes and behaviors, there is a paucity of solid research on the interrelation and influence of genetic and environmental factors on these parameters. The separate fields of psychology and molecular biology are subject to inherent limitations that may only be resolved through collaboration across disciplines. Certainly, areas relating to spirituality (Genospirituality) and political science are just two that are beginning to emerge as fruitful grounds for identification of specific polymorphic gene associations and may pave the way to advance a new science of human nature. We address the issue of *Nature vs. Nurture* as it relates to questions regarding the definition of happiness, its causes, and its promotion. These questions are central to understanding human nature and are emerging as an important target of research, especially in the area of nutrigenomics. The present commentary attempts to identify key "vector influences" that link genes, the brain, nutrition, and social behavior to the most desired, but misunderstood and potentially fragile experience known as *happiness*.

2.2.1 Genospirituality

There is an interesting new understanding of how certain genes may impact on human nature, our beliefs, and our spirituality. Certainly, this new understanding may have to do with how recovering addicts perceive the possible existence of a higher power. One even might argue that addiction *is* the attempt to utilize technologies to produce spiritual experiences. AA historian, Earnest Kurtz (2008)), suggested that one definition for addiction may be an attempt to use material reality to fill a spiritual void. The psychiatrist Carl Jung observed the pitfalls associated with this approach in a letter he wrote to Bill W. in 1961: "You see, Alcohol in Latin is 'spiritus' and you use the same word for the highest religious experience as well as for the most depraving poison. The helpful formula therefore is: *spiritus contra spiritum.*" The primary spiritual insight of AA is that one cannot *will* his/her way to recovery—an approach to life that led the alcoholic to futility and insanity. Rather, one must be *willing* to recover and learn to let go through the systematic program of the twelve steps. Thus, spirituality itself may be summarized as *letting go*.

The concept of *genospirituality* as elucidated by Charlton (2008) may provide insight into understanding how the acceptance of a higher power could restore sanity, as in Step 2. According to Charlton, the most frequently discussed role for genetic engineering is in relation to medicine. A second area that provokes discussion is the use of genetic engineering as an enhancement technology. Although there is increasing interest in the neuroscience of religion (McNamara 2009; Wildman 2011), one under studied area is the potential use of genetic engineering to increase human spiritual and religious experience—or genospirituality.

If technologies are devised that can conveniently and safely engineer genes causal of spiritual and religious behaviors, then people may become able to choose their degree of religiosity or spiritual sensitivity. For instance, it may become possible to increase the likelihood of direct religious experience, that is, *revelation*: the subjective experience of communication from the deity. Another potentially popular spiritual ability would probably be shamanism in which states of altered consciousness (e.g., trances, delirium, or dreams) are induced, and the shaman may undergo the experience of transformations, soul journeys, and contact with a spirit realm. Ideally, shamanistic consciousness could be modulated, such that trances were self-induced only when wanted and when it was safe and convenient and then switched-off again completely when full alertness and concentration are necessary.

Charlton continues: "It seems likely that there will be trade-offs for increased spirituality; such as people becoming less 'driven' to seek status and monetary rewards—as a result of being more spiritually fulfilled people might work less hard and take more leisure. On the other hand, it is also possible that highly moral, altruistic, peaceable and principled behaviors might become more prevalent; and the energy and joyousness of the best churches might spread and be strengthened. Overall, genospirituality would probably be used by people who were unable to have the kind of spiritual or religious experiences which they wanted (or perhaps even needed) in order to lead the kind of life to which they aspired."

Moreover, Nilsson et al. (2007) found that among boys, self-transcendence and spiritual acceptance were negatively correlated with the short 5-HTTLPR genotype and positively correlated with the short AP-2beta genotype. Among both boys and girls, significant interactive effects were found between 5-HTTLPR and AP-2beta genotypes, with regard to self-transcendence and spiritual acceptance. Boys and girls with the combination of presence of the short 5-HTTLPR and homozygosity for the long AP-2beta genotype scored significantly lower on self-transcendence and spiritual acceptance. In this regard, Comings and associates have found gene polymorphic associations with spirituality (Comings et al. 2000a, b). It is noteworthy that using a spiritual inventory as part of clinical history is important as an additional tool for medical treatment and diagnosis (Braverman 1987). Although controversial, clinical studies are beginning to clarify how spirituality and religion can contribute to the coping strategies of many patients with severe, chronic, and terminal conditions (Post et al. 2000). One interesting notion has received considerable attention; distinguishing religion from spirituality, especially in dying patients (Sulmasy 2006). Twin studies of spirituality showed that genes accounted for 50 % of the variance, the unique environment for 50 %, and the common environment, including cultural influences, for zero percent (Kirk et al. 1999).

This suggests that spirituality may be an intrinsic biological trait. By contrast, common environment and cultural transmission accounted for a significant percent of the variance of church attendance suggesting that religion is transmitted, at least in part, by non-genetic transmission from generation to generation (called *meme*) (Kirk et al. 1999). There are certain advantages that favor spirituality in terms of achieving well-being for both genders and these include but are not limited to the following: spirituality alleviates man's fears of his own death and of mortality; spirituality gives man control over a threatening world; near death experiences and spirituality; spirituality and optimism; spirituality and religion, and social cohesiveness; spirituality as a defense mechanism; inborn spirituality as a moral watchdog; and finding a spiritual mate.

The concept of spirituality having significant benefits for the human psyche, potentially leading to both optimism and happiness, may be independent of the existence or non-existence of God (Comings 2008). Hamer in his popular book "*The God Gene*" suggested that the selection for dopaminergic spirituality genes is driven by their ability to produce an innate sense of feel-good optimism akin to positive psychology. Accordingly, this would have selective value in the sense that optimism relates to the will to keep on living and procreating, despite the fact that death is inevitable (Comings 2008). Moreover, studies have shown that optimism seems to promote features that would have a positive selective value like better health and quicker recovery from disease (Braverman 1987). Newberg (2009) suggested a different kind of association of spirituality with a feel-good sensation. He suggested that the neurological machinery of spiritual transcendence may have arisen from the limbic system, which evolved for sex and mating. We find this very interesting, especially in light of work on dopaminergic genes and DA function, suggesting that DA is considered to be the pleasure and anti-stress molecule (Comings and Blum 2000). Newberg proclaimed, "It is no coincidence that mystics of all times and cultures have used the same expressive terms to describe their ineffable sexual experience: bliss, rapture, ecstasy and exaltation." He further suggested that the feel-good sensation was linked to the very neurological structures and pathways involved in transcendent experience, including the arousal, quiescent, and limbic jackpot! Finally, as in his provocative book "*Did Man Create God?*" Comings (2008) suggested that spirituality can be defined as a feeling of a connectedness with something greater than oneself "including any form of social order."

Perhaps, the greatest factor in the evolution of spirituality is that such a trait would maximize the development of man as a social animal. While the former concept is true, behavioral scientists suggest that the connection to a higher power is morphed by man's greatest fear, which is death. In consideration of one's spirituality, pondering the unattainable notion of our immortality and the doom of being mortal ultimately leads to feelings of doom with no escape at hand. We have witnessed it in the form of comedy as viewed in the film "*What About Bob?*" which explores the subject of "death therapy" and the never-ending question of inescapable death as observed by a prepubescent child. In this regard, Deepak Chopra in his book "*Life After Death*" (2006) clearly explained the many facets of people's beliefs as they relate to the possibilities of life after death. He emphatically asserted, "Those who have the least freedom of choice are driven by obsessions, compulsions, addictions, and unconscious impulses. To the extent that you become free of these, you have more choice. The same is true of a soul contemplating its next physical incarnation." According to Chopra, this is a very positive psyche attribute, especially to those who actually believe that death is not just dissolution and the end. However, in contrast to this belief, Comings in his book "*Did Man Create God*" (2008) pointed out that if consciousness is a prerequisite for the soul, and consciousness is extinguished when brain damage causes the loss of core consciousness, it would also cause the loss of the soul. According to Comings, "While the concept of a soul representing the essence of an individual and living on after death is central to many religions, its existence is not supported by modern neuroscience which states that consciousness, the spirit, and the soul are the products of neuronal activity and die when the person dies. This has major consequences for religion since without a soul there is no cosmic consciousness, no afterlife, no hell, no heaven, and thus no reward in heaven for good behavior." It is not the intent of this scientific treatise to address the existence or non-existence of God. It is, however, important to realize that the quality of and dependence on the cognizant connection to such a belief system can be significantly influential in an individual's ability to achieve a state of peace and happiness.

Other points of view may have this to say about this conjecture and it is these two independent views that have significant impact on one's happiness. According to one of us (JG), it is not the existence or non-existence of God that is at issue in this matter. It is an individual's state of mind that dictates their synaptic reality. It has been pointed out that this philosophical terrain is based on excessive psycho-religious speculation and psycho-neuro-genomic conjecture to explain how we created God. This is similar to determining whether other dimensions exist based on the mental competence of people. It exists or it does not exist. Whether or not we grasp it is irrelevant. We believe the key points can be made about the psycho-genomic mechanisms of how we connect to the concept of God and an afterlife without delving into atheistic hyperbole. The existence or non-existence of God is an entirely different issue than that of whether we do or do not have a conscious connection to a God-presence. As stated in the Big Book, "for faith in a Power greater than ourselves, and miraculous demonstrations of that power in human lives, are facts as old as man himself." To base the existence of God on a healthy state of brain chemistry is a very sophomoric premise, and dangerously eclipsing the boundaries and capabilities of science.

In terms of a few specific genes, Comings and associates (2000b) were the first to identify the role of a specific gene in spirituality. The gene was the DA D4 receptor gene. The gene was found to play a role in novelty seeking, one of the personality traits in Cloninger's Temperament and Character Inventory (Cloninger et al. 1993), and has been associated with compromised DA signaling in an in vitro study (Asghari et al. 1995). Comings and associates did identify genetic correlates of self-transcendence, but the association of this gene and novelty seeking linked to dopaminergic circuits did not emerge in a sample of substance abusers (Gelernter et al. 1997). Specifically, there was a borderline association with a self-forgetful sub-score but a strong association with spiritual acceptance. Other genes included the DA vesicular transporter gene (VMAT2), which was reported to be associated with spirituality. The fact that two different genes, DRD4 and VMAT2, have been found to associate with spirituality, and the fact that DA is the feel-good neurochemical, may help explain why spirituality plays a powerful role in the human condition and why the majority of people derive great comfort and happiness from a belief in a God. It is of further interest that in the Comings et al.'s study (2000b), those individuals who scored high on self-transcendence were less likely to abuse alcohol or drugs.

Accordingly, this may be because individuals whose reward pathways, and possibly other interacting pathways (e.g., serotonergic) activated by spirituality, would have less need to artificially activate their reward circuitry with foreign molecules like ethanol and cocaine. This is, indeed, the central pillar of AA's twelve steps (Gold and Dackis 1984). Moreover, Borg et al. (2003) at the Karolinska Institute in Sweden found that the binding of ethanol was lowest in those with the highest scores for self-transcendence, suggesting, "Such individuals had higher levels of brain serotonin." They showed that the HTRIA gene (5-HT1A receptor gene) was significantly associated with the self-transcendence scale and with spiritual acceptance (Borg et al. 2003). It is noteworthy that the lysergic moiety as in LSD is similar in structurally to serotonin having a modifying effects through psychedelic spiritual experiences (see preamble note on Bill W).

Finally, many different plants around the world contain a range of psychedelic drugs (serotonergic, opioid, and catecholaminergic), which are capable of strongly enhancing man's spirituality and/or spiritual awareness. It is suggested that this is accomplished by providing a powerful feeling of communication with a supernatural power (Shultes et al. 1998). Comings (2008) further points out that these entheogens (good-producing substances) played a profound and critical role in facilitating human's early belief in a god or gods and in the development of religion.

Whatever one's belief, the twelve steps are extremely important for providing a positive attitude about an individual's addiction/reward dependence behavior. Understanding the role of genes with regard to spirituality while in its infancy should lead to development of potential therapeutic targets in the future because there is a powerful genetic linkage between humans and God.

2.2.2 Sanity

In the Big Book, Bill W referred to insanity as, **"… a lack of proportion, of the ability to think straight…."** From an AA perspective, insanity includes a recurring denial such that the same decision has to be made over and over because the same lack of evidence about losses from drinking keeps recurring. Thus, the alcoholic does the same harmful thing again and again.

By comparison, other meanings of the word sanity are simplistic: The quality or condition of being sane; soundness of judgment or reason (Middle English sanite, *health*, from Old French, from Latin sanitas, from sanus, *healthy*).

> Sanity is the lot of those who are most obtuse, for lucidity destroys one's equilibrium: it is unhealthy to honestly endure the labors of the mind which incessantly contradict what they have just established (Georges Bataille).

> No man is sane who does not know how to be insane on proper occasions (Henry Ward Beecher).

> What frightens us most in a madman is his sane conversation (Anatole France).

> Show me a sane man and I will cure him for you (Carl Jung).

> In a mad world, only the mad are sane (Akiro Kurosawa).

> What sane person could live in this world and not be crazy? (Ursula K. LeGuin).

Alfred Korzybski proposed a theory of sanity as early as 1933. He believed that sanity was tied either to the structural fit of the world or to lack of any structural fit that is actually going on in the world. Psychiatrist Phillip S. Graven suggested the term *un-sane* to describe a condition that is not exactly insane, but not quite sane either (Korzybski 1994).

In "*The Sane Society,*" published in 1955, psychologist Erich Fromm proposed that not just individuals but entire societies "may be lacking in sanity." Fromm argued that one of the most deceptive features of social life involves "consensual validation" (Fromm 1955).

According to Fromm,

> It is naively assumed that the fact that the majority of people share certain ideas or feelings proves the validity of these ideas and feelings. The fact that millions of people share the same vices does not make these vices virtues, the fact that they share so many errors does not make the errors to be truths, and the fact that millions of people share the same form of mental pathology does not make these people sane.

From the perspective of AA, it is not necessary to apply clinical standards to the constructs of sanity and insanity. Just as an individual may qualify as an alcoholic on the experiential basis of not "staying stopped," so too may he or she qualify as insane on the basis of expecting different results from the same behavior. More to the point, these are determinations (alcoholic and insane) that individuals make about themselves on the basis of identification with someone else's experiences. Validating or invalidating these self-appraisals according to a clinical model is irrelevant, as far as AA is concerned. What matters is what the individual believes based on experience. For the same reason, theological and scientific arguments for or against the existence of God are irrelevant to confirming the pragmatic value of *belief* in God, which may be amply verified by the experience of many AA members.

For the present treatise, we the authors take the liberty of discussing human sanity in the same contextual frame of executive function and judgment. Sanity of course is more complicated, and we advise caution related to our quest to tie the second step to brain biology and genetics. Our group recently published a paper on the neuropharmacology and neurogenetics aspects of executive functioning (Bowirrat et al. 2012). Accordingly, executive functions are processes that act in harmony to control behaviors necessary for maintaining focus and achieving outcomes (sanity). Executive dysfunction in neuropsychiatric disorders is attributed to structural or functional pathology of brain networks involving prefrontal cortex and its connections with other brain regions. The prefrontal cortex receives innervations from different neurons associated with a number of neurotransmitters, especially DA. Bowirrat et al. (2012) suggested that examination of multifactorial interactions of an individual's genetic history, along with environmental risk factors, can assist in the characterization of executive functioning (includes working memory and judgment) for that individual. Based upon the results of genetic studies, they also proposed genetic mapping as a probable diagnostic tool serving as a therapeutic adjunct for augmenting executive functioning capabilities. It is concluded that preservation of the neurological underpinnings of executive functions requires the integrity of complex neural systems including the influence of specific genes and associated polymorphisms to provide adequate neurotransmission. While this is a very complex area of research, we will restrict our remarks to neurogenetics of the Brain Reward Cascade (Blum and Kozlowski 1990) and executive function (see Fig. 1.2).

Bertolino et al. (2010) demonstrated that a functional SNP (rs1076560) within the DA D2 receptor gene (DRD2) predicted striatal binding of the two radiotracers to DA transporters and D2 receptors, as well as the correlation between striatal D2 signaling with prefrontal cortex activity during performance of a working memory task. These findings are consistent with the possibility that the balance of excitatory/inhibitory modulation of striatal neurons may also affect striatal outputs in relationship with prefrontal activity during working memory performance within the cortico-striatal-thalamic-cortical pathway. Furthermore, Markett et al. (2010) found a significant interaction between nicotinic acetylcholine receptor (rs#1044396) and a haplotype block covering all three dopaminergic polymorphisms (rs#1800497, rs#6277, rs#2283265) on working memory capacity. This effect only became apparent on higher levels of working memory load. This is the first evidence from a molecular genetics perspective that these two neurotransmitter systems interact on cognitive functioning. Frank and Hutchinson (2009) found effects of the commonly studied Taq1A polymorphism on reinforcement-based decisions were due to indirect association with C957T of the DA receptor gene.

It is noteworthy that overexpression of D2 under pathological conditions such as schizophrenia could give rise to motivational and timing deficits (Drew et al. 2007). The increase in DA D2 receptors has been shown by others to prevent storage of lasting memory traces in prefrontal cortex networks and impair executive functions (Xu et al. 2009).

Interestingly, Stelzel et al. (2009) showed that COMT Val158Met polymorphism effects on working memory performance are modulated by the DRD2/ANKK1-TAQ-1A polymorphism. Val − participants, characterized by increased prefrontal DA concentrations, outperformed Val + participants in the manipulation of working memory contents, but only when D2 receptor density could be considered to be high. Stelzel et al. (2009) suggested that this beneficial effect of a balance between prefrontal DA availability and D2 receptor density reveals the importance of considering epistatic effects and different working memory sub-processes in genetic association studies. However, these genetic effects may not be present or are too subtle to detect in healthy subjects (Blanchard et al. 2011). Moreover, it has also recently been found that participants with Val/Val of the COMT gene involved in a 17-week running exercise program improved cognitive performance to a greater extent compared with individuals carrying a Met allele. From these results, it was concluded that an increase in physical fitness provided a means to improve cognitive functioning via dopaminergic modulation (Stroth et al. 2010). Certainly, in an attempt to get back to sanity, many recovery addicts develop fruitful exercise programs. In a related study of HIV and methamphetamine dependence, dopaminergic overactivity in prefrontal cortex conferred by the Met/Met genotype appeared to result in a liability for executive dysfunction (poor judgment) and potentially associated risky sexual behavior (Bousman et al. 2010).

To reiterate, the dopaminergic system of the brain is heterogeneous and multifunctional. It is a system with many important neurochemical functions and has been credited with resultant behavioral effects such as "pleasure," "stress reduction," and "wanting" (Blum et al. 1997, 2000; Bowirrat and Oscar-Berman 2005; Blum et al. 2008). In addition to its contributory role in addressing higher-order cognition and executive functions, it is very important in relapse of substance-seeking behavior (Blum et al. 2009b; Blum et al. 2011c). It is well known that acute alcohol intoxication and long-term chronic alcoholism significantly impair executive functioning and judgment (Oscar-Berman and Marinkovic 2007). Lyvers and Tobias-Webb (2010) found a dose-dependent selective disruption of prefrontal cortex functioning by alcohol. They suggested that alcohol-associated perseveration on the Wisconsin Card Sorting Test may reflect the inhibitory effect of alcohol preventing DA release in the prefrontal cortex.

Finally, being somewhat complex, abnormal increases in D2 receptor activity cause a more general impairment in behavioral flexibility especially in patients with attention deficit hyperactivity disorder (Barnes et al. 2011). These findings suggest that deficits in these forms of executive functioning observed in disorders linked to dysfunction of the DA system may be attributable in part to aberrant increases or decreases in mesoaccumbens DA activity (Zang et al. 2007; Arnsten 2009; Dickinson and Elvevåg 2009; Haluk and Floresco 2009; MacDonald et al. 2009; Braet et al. 2011). The control of DA release in prefrontal cortex and other brain regions is regulated by many neurotransmitters and second messenger genes that constitute a genetic map that could provide important information relating to a predisposition to poor judgment.

2.2.3 Development of a Genetic Map to Identify Individuals at Risk for Impaired Judgment

Executive dysfunctions are linked to flawed DA metabolism and especially to low D2 receptor density as well as other neurotransmitter deficits due to specific gene polymorphisms. Moreover, executive dysfunctions result from abnormalities in the corticomesolimbic system of the brain. These directly link abnormal craving behavior with a defect in the DRD2 gene and with other dopaminergic genes (D1, D3, D4, and D5, DATA1, MAO, COMT), including many genes associated with the brain reward function (Pinto and Ansseau 2009; Pinto et al. 2009).

The genesis of all behavior, be it normal (socially acceptable) or abnormal (socially unacceptable), derives from an individual's genetic makeup at birth. This relates to the concept of Dr. William D. Silkworth (*The Doctor's Opinion* in the *Big Book*) that the action of alcohol on alcoholics is a manifestation of an "allergy" (a phenomenon of craving) that does not occur in most temperate drinkers.

This genetic predisposition, due to multiple gene combinations and polymorphisms, is expressed differently based on numerous epigenetic processes of neurological adaptation through experience and environmental factors. These factors include family, friends, educational and socioeconomic status, environmental contaminant exposure, and the availability of psychoactive drugs and unhealthy foods. The core of predisposition to these behaviors is a set of genes interacting with the environment, which promote a feeling of well-being via neurotransmitter interaction at the reward center of the brain (located in the mesolimbic system) leading to normal DA release (Blum et al. 1996a; Hack et al. 2011).

Subjects afflicted with executive decision-making dysfunction carry polymorphic genes in dopaminergic pathways that result in hypodopaminergic function caused by a reduced number of DA D2 receptors, reduced synthesis of DA (by DA beta–hydroxylase), reduced net release of presynaptic DA (from, for example, the DA D1 receptor), increased synaptic clearance due to a high number of DA transporter sites (DA transporter), and low D2 receptor densities (DA D2 receptor), making such people more vulnerable to addictive behaviors and relapse because of poor judgment (Comings and Blum 2000; Blum et al. 2011c).

The inability to make good decisions (sanity) involves shared genes and their mRNA expressions and behavioral tendencies, including dependence on alcohol, psychostimulants, marijuana, nicotine (smoking), opiates, altered opiate receptor function, carbohydrate issues (e.g., sugar binging), obesity, pathological gambling, premeditated aggression, stress, pathological aggression, and certain personality disorders, including novelty seeking, and sex addiction.

The common theme across all of these substances and behaviors is that they induce presynaptic DA release (Dreyer 2010). Spectrum disorders such as ADHD, Tourette's syndrome, and autism are also included due to DA dysregulation. Other rare mutations (Sundaram et al. 2010) have been associated with Tourette's and autism. One example includes the association with Neuroligin 4 (NLGN4), a member of a cell adhesion protein family that appears to play a role in the maturation and function of neuronal synapses (Lawson-Yuen et al. 2008) that could have an impact on executive functions. The roles of many of these reward genes and associated polymorphisms have been reviewed in recent papers by Blum et al. (2012a, b).

Sanity (sound judgment) or insanity (repetitive behavior in spite of harm) may be impaired even at birth and could be due to deficient brain reward circuitry function, especially resulting in a hypodopaminergic trait. This poor judgment could be a root cause for aberrant substance-seeking behavior in the face of harm's way. This becomes further complicated when other environmental factors are present including drug availability, non-nurturing parents, social economic burdens and stress. Importantly, the ability to behave sanely also may be impacted by an individual's relationship with a power greater than themselves. In terms of relapse, it is well known that the prefrontal cortex and cingulate gyrus are very important brain regions that could regulate relapse. Poor judgment stemming from impairments in the neurochemical functioning of these regions due to genes and/or toxic substances and/or behaviors impedes recovery and induces relapse. Understanding the molecular biology of the brain reward system (genes and environment) highlights the importance of positive input from fellowship (self-help) programs and other treatment modalities that can offset unwanted gene expression, lift spirits, and assist in enabling the individual to achieve a state of sanity and make good choices.

2.3 Step 3: Made a Decision to Turn Our Will and Our Lives Over to the Care of God as We Understood Him

2.3.1 Will/Control

The third step discusses the fact that the individuals have decided to turn their will/control over to the care of God.

The distinction between will and control is essential to understanding the third step. The control the alcoholic is supposedly "out of" refers primarily to his/her involvement with the substance of alcohol. The *Big Book* states, "We alcoholics are men and women who have lost the ability to control our drinking" (*Big Book*, p. 30). The real culprits behind this state of affairs, according the AA literature, are the "phenomenon of craving" that occurs while drinking and the "strange mental blank spots" that occur when presented with the opportunity to drink. Hence, there are bodily and mental components to the "illness" of alcoholism. The third component of this illness is the spiritual element of the individual. The culprit underlying the spiritual malady is described as follows: "Selfishness—self-centeredness! That, we think, is the root of our troubles." The "insanity" from which the alcoholic seeks restoration (Step 2) is related as much to his/her "powerlessness over alcohol" as it is to his/her life being "unmanageable" (Step 1). The *Big Book* suggests that the unmanageability of life is a result of the individual imposing his/her will (purpose, end, or design) onto the people, places, and things around him or her: "So our troubles, we think, are basically of our own making. They arise out of ourselves and the alcoholic is an extreme example of self-will run riot, though he usually doesn't think so" (*Big Book*, p. 62). Sanity emerges, on this view, from living in accord with some "higher purpose" for one's life. "Established on such a footing we became less and less interested in ourselves, our little plans and designs. More and more we became interested in seeing what we could contribute to life" (*Big Book*, p. 63). Thus, the "will" and "willpower" are also different. The "will" refers to the organizing purpose of one's life. Turning the "will" over, then, denotes an interior process of *letting go* of one's selfish purposes by means of a decision. "Willpower," however, refers to the strength of the individual's resolution to abstain. Within twelve-step culture, exercising willpower against a drug of choice is likened to holding one's breath. Eventually, you are going to need to breathe (Thompson, personal communication).

We believe that understanding the neurochemical basis of alcohol/drug-seeking behavior will provide support for the decision to turn the addicts' will/control over to the care of God. This includes understanding the efforts to develop non-human animal models for the study of neurobiological mechanisms underlying chronic alcohol drinking, alcoholism, abnormal alcohol-seeking behavior, as well as drug and food seeking. Selective breeding has produced stable lines of rodents that reliably exhibit high and low voluntary alcohol consumption (McClearn 1972).

In addition, according to McBride and Li (1998), models of chronic ethanol self-administration have been developed in rodents without a genetic predisposition for high alcohol-seeking behavior, to explore environmental influences in ethanol drinking and the effects of physical dependence on alcohol self-administration. The selectively bred high-preference animals reliably self-administer ethanol by free-choice drinking and operantly responded for oral ethanol in amounts that produce pharmacologically meaningful blood alcohol concentrations (50–200 mg % and higher). In addition, the alcohol-preferring rats will self-administer ethanol by intragastric infusion. With chronic free-choice drinking, the high alcohol-preferring rats develop tolerance to the high-dose effects of ethanol and show signs of physical dependence after the withdrawal of alcohol. Compared with non-preferring animals, the alcohol-preferring rats are less sensitive to the sedative–hypnotic effects of ethanol and develop tolerance more quickly to high-dose ethanol.

Non-selected common stock rats can be trained to chronically self-administer *ethanol* following its initial presentation in a palatable sucrose or saccharin solution, and the gradual replacement of the sucrose or saccharin with *ethanol* (the sucrose/saccharin-fade technique). Moreover, rats that are trained in this manner and then made dependent by ethanol vapor inhalation or liquid diet increase their ethanol self-administration during the withdrawal period. Both the selectively bred rats and the common stock rats demonstrate "relapse" and an alcohol deprivation effect following two or more weeks of abstinence. This finding is very important because it points out that even individuals without a genetic deficit in brain reward chemistry will, under chronic intake of alcohol, display relapse tendencies against their will.

Moreover, systemic administration of agents that (1) increase synaptic levels of 5-HT DA, (2) activate 5-HT1A, 5-HT2, D2, D3 receptors, or (3) block GABA(A) and 5-HT3 receptors decrease ethanol intake in most animal models. Neurochemical, neuroanatomical, and neuropharmacological studies indicate innate differences exist between the high alcohol-consuming and low alcohol-consuming rodents in various brain limbic structures. In addition, reduced mesolimbic DA and

5-HT functions have been observed during alcohol withdrawal in common stock rats. Depending on the animal model under study, abnormalities in the mesolimbic DA pathway, and/or the 5-HT, opioid, and GABA systems that regulate this pathway may underlie vulnerability to the abnormal alcohol-seeking behavior in the genetic animal models (McBride 1998).

In terms of stress-induced alcoholism, our laboratory developed a schema to understand the relationship between genetics and environmental etiological indices for drug-craving behavior (inability to stop). Consensus of the literature points toward a neuropsychogenetic model of alcoholism. Evidence in both animals and humans tends to support the proposed "genotype" theory of alcohol-seeking behavior, whereby a predisposition to alcohol preference may be mediated in part by either innate (genetic) or environmentally (stress and/or alcohol) induced brain opioid peptide dysfunction (Blum and Topel 1986; Blum et al. 1986).

Scrutiny of the data from a series of studies performed by Blum's group reveals that the C58/J alcohol-preferring mice have significantly lower baseline methionine-enkephalin levels in both the corpus striatum and the hypothalamus compared to C3H/CHRGL/2 non-alcohol-preferring mice. In other brain regions in these two strains, specifically pituitary, amygdala, midbrain, and hippocampus, analyses of methionine-enkephalin levels did not show any significant differences. This suggests that the hypothalamus may indeed be a specific locus involved in the regulation of alcohol intake, via the molecular interaction between neuro-amines and opioid peptides, as they are influenced by genetics and environment (Blum et al. 1987b).

Additionally, Blum et al. (1982b) investigated ethanol preference utilizing a three-choice two-bottle preference test with both the C57Bl/6 N (alcohol non-preferring) and C57Bl/6 J (alcohol preferring) mice. In addition, these sublines were evaluated for whole brain methionine-enkephalin levels, which were significantly lower in C57Bl/6 J mice compared to C57Bl/6 N mice. This finding supports the involvement of the peptidyl opiates in ethanol-seeking behavior (Blum et al. 1982b).

This work was followed up by showing a negative correlation between the amount of ethanol (10 %) consumed and endogenous levels of brain [Met]-enkephalin. Specifically, in C57BL/mice after 1-day starved groups of both C57BL/6 J and DBA/2 J mice were challenged with ethanol (10 %) for a 1-day acceptance of ethanol test, The C57BL/6 J animals, having significantly lower levels of brain [Met]-enkephalin compared with their non-alcohol-treated controls, accepted higher amounts of alcohol than DBA/2 J controls. These results further suggest that the brain endogenous peptidyl opiates may play a crucial role in alcohol-seeking behavior (Blum 1983).

These experiments led Blum's group to explore the potential therapeutic rationale involving the utilization of novel inhibitors of carboxypeptidase-A (enkephalinase) which raises endogenous enkephalin levels and possess anti-alcohol-seeking effects (Blum et al. 1987a). In fact, they have been able to significantly attenuate both volitional and forced ethanol intake, respectively, by acute and chronic treatment with hydrocinnamic acid (metabolite of D-phenylalanine) and D-phenylalanine, known carboxypeptidase (enkephalinase) inhibitors. Since these agents, through their enkephalinase inhibitory activity, raise brain enkephalin levels, Blum et al. (1987a) proposed that excessive alcohol intake can be regulated by alteration of endogenous brain opioid peptides. This finding may have great relevance as a "pharmacogenetic engendering" tactic to block unwanted (against will) craving behavior for alcohol, drugs, and even food (Blum et al. 1987a).

The question of nature versus nurture in terms of substance-seeking behavior has been addressed in many studies across the globe. We must be cognizant that it is always the relationship between both the environment and our genome that predicts any behavioral outcome: $P = G + E$.

Recently, Campbell et al. (2009) reviewed problematic aspects of alcohol abuse disorder (excessive alcohol consumption) and the inability to refrain from alcohol consumption during attempted abstinence (loss of willpower during recovery). While being cognizant of the significant role of genetics (up to 70 % contribution of the variance), they concluded that early environmental trauma alters neurodevelopmental trajectories that can predispose an individual to substance abuse and a numerous neuropsychiatric disorders.

Prenatal stress is a well-established protocol that produces perturbations in nervous system development, resulting in behavioral alterations that include hyper-responsiveness to stress, novelty, and psychomotor stimulant drugs (e.g., cocaine, amphetamine). A great body of information exists to suggest that depression in female offspring continues through adulthood. This is related to a decrease in dopaminergic neurotransmission at the NAc, and the present provides new evidence in support of the hypothesis that maternal stress during gestation increases the risk of depression in the offspring (Alonso et al. 1994).

Moreover, prenatally stressed animals exhibit enduring alterations in basal and cocaine-induced changes of DA and glutamate transmission within limbic structures, which exhibit pathology in drug addiction and alcoholism, suggesting that these alterations may contribute to an increased propensity to self-administer large amounts of drugs of abuse or to relapse after periods of drug withdrawal. Given that cocaine and alcohol have actions on common limbic neural substrates (albeit by different mechanisms), it has been hypothesized that prenatal stress would elevate the motivation for, and consumption of, alcohol.

Accordingly, Campbell et al. (2009) found that male C57BL/6 J mice subject to prenatal stress exhibited higher operant responding and consumed more alcohol during alcohol reinforcement as adults. Alterations in glutamate and DA neurotransmission within the forebrain structures appeared to contribute to the prenatal stress-induced predisposition to high alcohol intake and were induced by excessive alcohol intake (Campbell et al. 2009).

Willpower is not simple to control, especially if you are born with a compromised reward system, especially low levels of endorphins. Genetically predisposed individuals seek out drugs such as alcohol, heroin, cocaine, nicotine, and even sugar, because these substances all activate reward substrates (i.e., enkephalins, DA pathways) and provide a pseudo temporary feeling of well-being (so-called normalization). Willpower is based on both the interplay of genes and environmental elements in society. This includes stress as an adult and surprising during the prenatal phase. This early stress could lead to aberrant substance use disorders in adult life. Since it is not easy to fight the hard wiring of our brain reward circuitry, for the recovering addict, it seems obvious to look for reward outside of our genome (i.e., alcohol, drugs, sex, and food).

2.4 Step 4: Made a Searching and Fearless Moral Inventory of Ourselves

The fourth step involves a systematic process of assessing the "causes and conditions" that "drove [the individual] to drink." It extends somewhat naturally from the third step, which involved "a decision to turn our life and will over to God *as we understood Him*." The fourth step is the first practical action related to that decision. This step does not entail the enumeration of the individual's deficiencies. Rather, it designates "resentment as the 'number one' offender" in terms of the selfishness that is supposedly indicative of the alcoholic's spiritual illness. The *Big Book* describes a process of the individual making a list of all his/her resentments against "people, principles, or institutions"; examining the manifestations of self that these resentments affected, such as self-esteem, pride, financial or emotional security, personal or sexual relations; then examining the role that the individual making the list might have played in the production of each resentment, including how he or she was being selfish, dishonest, self-seeking, or afraid. In this way, the individual takes a step in the direction of letting go of his or her self—that is, his or her life and will. (This is the list that is read or "admitted" to another human being and to God in Step 5.)

The first consideration that may be helpful to the recovering individual is to realize that addiction to alcohol *per se* may not be the sole culprit. In terms of taking an inventory of ourselves, it is very easy to dismiss associated issues such as anger, the inability to cope with stress, to focus, to complete tasks, a poor job history, antisocial issue, overeating, oversexing, gambling, gaming on the Internet, polygamous relationships, high risk taking, lying, cheating, criminal instincts, smoking, getting high, or being unhappy with oneself and life (suicidal ideation).

At a NA meeting a man stated, "In my own head I am not a good man."

The first consideration has been the subject of intense investigation and involves the concept of *Reward Deficiency Syndrome* (Blum et al. 1996b). The human brain produces a number of "feel-good" chemicals (e.g., DA, norepinephrine, GABA, and endorphins), all of which work together to produce feelings of well-being (Blum and Kozlowski 1990). When levels of these chemicals are low or blocked from the brain's receptors, pain, discomfort, and agitation are the result. This "reward deficiency" is associated with mood instability, anxiety, hypersensitivity, and irritability. Individuals with a family history of alcoholism, other addictions, Tourette's syndrome, or attention deficient disorder may be born with an inability to produce or utilize feel-good chemicals in the same way as people who do not have this genetic makeup. Specifically, while multiple polygenes are involved in any phenotype (behavior), defects in the D2A1 allele, a DA receptor gene, have been linked to these disorders (Blum et al. 1990, 1991). People who have this defective gene lack a sufficient number of DA receptors (30–40 % less) in their brains to produce the normal *neurochemical reward cascade*, creating Reward Deficiency Syndrome. Exposure to prolonged periods of stress and alcohol or other toxic substances can also lead to a corruption of the cascade function. It is often quite difficult to determine, however, which came first—the phenotypic defect or the defect to the gene. Understanding this concept will provide the basis for transfer of addictions from, for example, alcohol to coffee to sugar to gambling and sex. Recently, Klavis and Brady (2012) addressed this issue in a paper where they suggested that based on neuroscience research we may be closer to getting to the core of addiction by hatching the addiction egg possibly due to dopaminergic genetics.

It should be noted that this explanation may be sufficient for understanding the transfer of addictions from a neurobiological viewpoint, but the liberal application of the twelve steps to virtually any other addiction indicates that the other fellowships have found something useful in AA's approach to understanding the root cause of addiction in terms of the frustration and futility that stem from trying and failing to impose one's will on to the people, places, and things around him or her.

In any case, those suffering from Reward Deficiency Syndrome are unable to produce a feeling of well-being and consequently often self-medicate with substances or behaviors that help raise the levels of feel-good chemicals in their system, even if only temporarily (Di Chiara and Imperato 1988). They feel good and function better with less angst, agitation, and emotional pain. These substances often include alcohol, nicotine, other stimulants, illicit drugs, junk foods, sugars, carbohydrates, or thrill-seeking behaviors such as gambling, sex, and Internet porn. Unfortunately, this only provides temporary relief while bringing with it the possibility of more long-term consequences (http://www.livestrong.com/article/15011-what-is-the-reward-deficiency-syndrome/#ixzz1sjAYqhWW).

A second consideration is at what point could a recovering dependent human truly take a fearless moral inventory of oneself? If one's neural network is compromised not only by the taking of substances or excessive behaviors (process addictions) but also by his/her DNA, this is almost impossible. In other words, if the brain is compromised either from drugs or from genes how then could we expect addicts to actually achieve this seemingly important step in the recovery process?

2.4.1 Brain Reward Circuitry Impairment During Protracted Abstinence

According to Volkow and her colleagues, drug addiction is characterized by a compulsive drive to take drugs despite serious negative consequences, and it is a disorder that involves complex interactions between genetic (Volkow et al. 2006) and environmental variables (Volkow et al. 2012a, b). For example, undoubtedly heroin addiction is a complex phenomenon of the brain, involving both affective and cognitive process (Volkow et al. 2003). It has been found that in heroin-dependent individuals, compared to healthy subjects, there is increased white matter intensity in the frontal area and decreased gray matter density in prefrontal cortex and in regions of the temporal lobes (Yuan et al. 2009; Zhang et al. 2011). Zang et al. (2011) also found a high accuracy in the activation pattern differences between heroin-dependent subjects and healthy individuals during resting-state brain activities (Zang et al. 2011). These differences of activation patterns included: orbitofrontal cortex, the hippocampal/parahippocampal region, amygdala, caudate, putamen, insula, and thalamus (Bell et al. 2011; Wang et al. 2011a, b).

Importantly, Volkow et al. (2003) proposed a map consisting of four circuits involved in drug abuse and possibly reward behaviors (i.e., addiction(s): (1) reward, located in the NAc and ventral pallidum; (2) motivation/drive, located in the orbitofrontal cortex and subcallosal cortex; (3) memory and learning, located in the hippocampus and amygdala; and (4) control, located in the prefrontal cortex and anterior cingulate gyrus. Our current knowledge indicates that whereas aberrant craving behavior resides in the caudate-accumbens brain region, loss of control and thus relapse occurs in the cingulate gyrus (Goldstein et al. 2004). Moreover, Thanos et al. (2008) as well as Rothman et al. (2007) independently suggested that DA agonist therapy by either increasing D2R availability and or enhanced DA release, respectively, could be useful therapeutic adjuncts for the treatment of cocaine and alcohol addictions, as well as for obesity, attention deficit disorder, and depression, or RDS behaviors (Blum et al. 1996b).

Previously, Liu's group found resting-state functional abnormalities in heroin-dependent individuals affecting brain functional organization (Yuan et al. 2009; Zhang et al. 2011). These functional impairments could negatively impact decision-making and inhibitory control. Moreover, they found in earlier studies that, compared to normal controls, heroin addicts showed reduced activation in right amygdala in response to the affective pictures, consistent with previous reports of blunted subjective experience for affective stimuli in addicts. Other studies showed persistent abnormalities in orbitofrontal cortex function following one month of heroin withdrawal. This point must be considered when treating a heroin-dependent patient and asking the individual to take inventory of self (Wang et al. 2011b). Usually, however, a prolonged period of abstinence is suggested by AA before attempting to take the fourth step, as this is typically sufficient for the individual to "dry out" enough to embark on self-appraisal.

Zijlstra et al. (2008) found lower baseline D2R availability in opiate-dependent subjects than controls in the left caudate nucleus. D2R availability in the putamen correlated negatively with years of opiate use. For people in Narcotics Anonymous previously addicted to heroin, it may take years to have enough clarity to successfully accomplish Step 4.

Moreover, Zijlstra's group (2008) further found that after cue-exposure, opiate-dependent subjects demonstrated higher DA release than controls in the right putamen. Chronic craving and anhedonia were positively correlated with DA release.

Treatment strategies that increase D2Rs, therefore, may be an interesting approach to prevent relapse in opiate addiction. To this aim, we evaluated the role of KB220Z™ on reward circuitry in a triple blinded-randomized placebo controlled cross-over study in five heroin addicts undergoing protracted abstinence (average 16.9 months). Based upon unpublished data summarized and presented at the World Congress of Psychiatric Genetics (Washington, DC, 2011), Blum reported that KB220Z ™ induced a BOLD activation of caudate-accumbens dopaminergic pathways compared to placebo following 1-h acute administration. Furthermore, KB220Z also reduced the higher dopaminergic activity in the putamen (see Fig. 2.2). Moreover, in 10 heroin-dependent subjects, we found three brain regions of interest to be significantly activated from resting state ($p < 0.05$) (see review by Blum et al. 2012a, b). These results coupled with other qEEG studies of this compound suggest a putative anti-craving/anti-relapse role in drug addicts by direct or indirect dopaminergic interaction (Blum et al. 2010; Miller et al. 2010) (see Fig. 2.3).

Our present findings derived from this small pilot study showing a clear difference between placebo and KB220Z™ in terms of BOLD activation of the dopaminergic pathways of the caudate-accumbens area is encouraging. Due to a small number, we cannot determine statistical significance and as such the

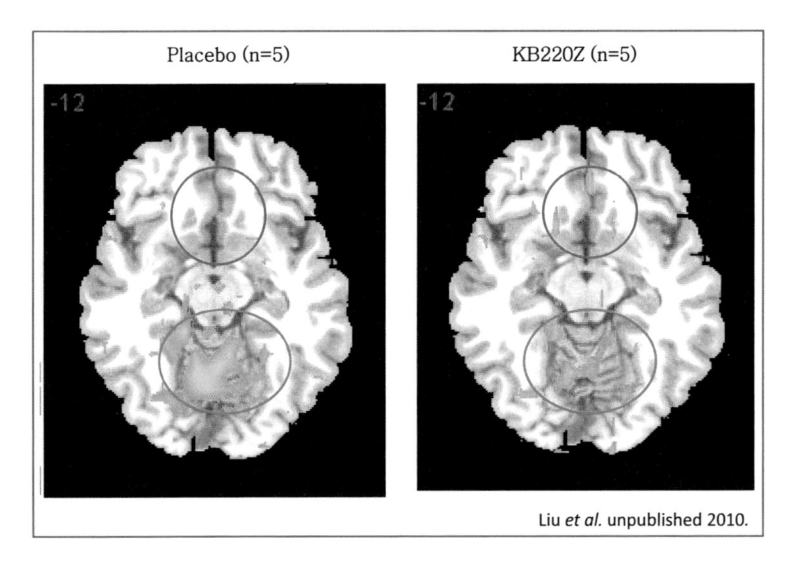

Fig. 2.2 Resting-State fMRI After One Dose KB200Z versus placebo. The fMRI study compared KB220Z to placebo, 1 h after dosing, in abstinent heroin-dependent patients in China. This 2 × 2 design experiment shows the resting state of the fMRI scan of the same five protracted abstinent heroin addicts 1 h after receiving an acute dose of placebo or KB220Z. The scan represents the effect of an acute dose of KB220Z on the caudate-accumbens-putamen brain region. Notice in the *orange circles* that there is a strong activation of the dopamine reward site. Moreover, KB22Z induced a "normalization" of the putamen region—*blue circles*. It is hypothesized that KB220Z caused dopaminergic agonistic activation of dopamine D2 receptors promoting enhanced reward and normalization (Modified: Liu et al. [unpublished]; Blum et al. 2013 with Permission)

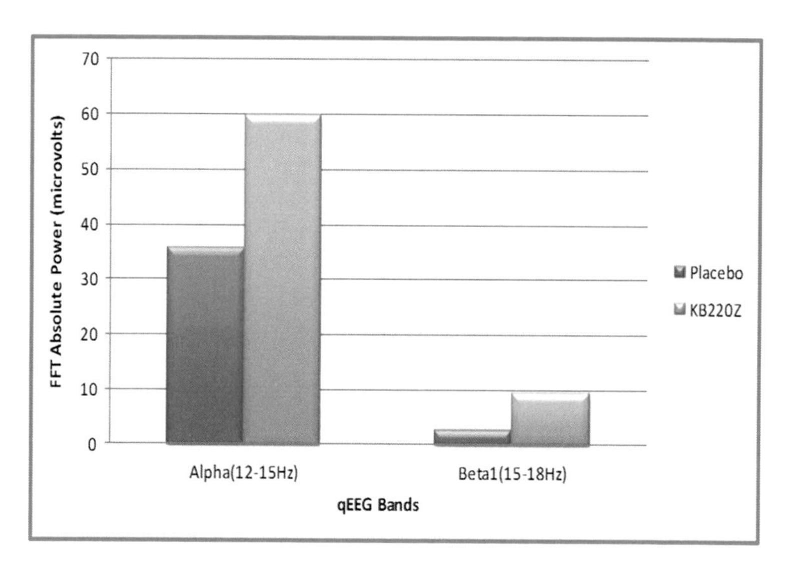

Fig. 2.3 KB200Z normalizes qEEG dysregulation. In a 2 × 2 cross-over design 1 h after administration of KB220Z, the FFT Absolute Power Alpha and Low Beta Bands increased compared to placebo in 10 psychostimulant-dependent abstinent subjects (Modified with permission Blum et al. 2010)

experiment will continue by adding additional heroin-dependent subjects. However, it is important that we did find statistical significance ($p < .05$) when we evaluated the 10 subjects at rest compared to KB220Z™ treatment in three important brain regions of interest. We are further encouraged that in this pilot study, we did observe an attenuation of the resultant hyperactivity in the putamen of abstinent heroin-dependent subjects. Currently, albeit knowing that there is a lower D2R availability in the putamen of abstinent heroin-dependent subjects, we do not understand the mechanism by which KB220Z™ administration (after 1 h) induced an attenuation of this hypo-state. This will be the subject of additional investigation, and it may involve white matter abnormalities (Blum et al. 2010b).

In ongoing research, we also are exploring the role of KB220Z compared to placebo in cue-induced craving behavior as well as the effect of KB220Z on white matter synapses. This additional experiment is important since the structure and function of white matter synapses become increasingly relevant in disease. White vesicular neurotransmitter release has been known to be the preserve of gray matter; it is known that synaptic style release of glutamate occurs deep in white matter. As white matter is increasingly well recognized as a substrate for disease, dysregulation of white matter synaptic transmission will play a role in Reward Deficiency Syndrome and a number of impulsive/compulsive/addictive behaviors.

Ruiz et al. (2013) used MRI to examine white matter volumes in alcoholics and controls and observed group differences primarily in the corpus callosum. Years of heavy drinking had a strong negative impact on frontal and temporal white matter among alcoholic women and on the corpus callosum among alcoholic men.

Quantity of alcohol consumption was associated with smaller corpus callosum and larger ventricular volumes among alcoholic women, while abstinence duration was associated with larger corpus callosum volume among alcoholic men. While duration of heavy drinking was positively associated with age, multiple regression analyses showed that duration of heavy drinking was a stronger predictor of lower white matter volumes than age in many regions.

Using diffusion tensor neuroimaging, Harris et al. (2008) found that compared to non-alcoholic controls, abstinent alcoholics had diminished frontal lobe fractional anisotropy (a measure of axonal integrity) in the right superior longitudinal fascicles II and III, orbitofrontal cortex white matter, and cingulum bundle. These fractional anisotropy measures provided 97 % correct group discrimination. Working memory scores positively correlated with fractional anisotropy measures in the superior longitudinal fascicle III in control subjects only. The findings demonstrated white matter microstructure deficits, which may contribute to underlying dysfunction in memory, emotion, and reward response in alcoholism.

Interestingly, current cocaine-dependent users show reductions in white matter integrity, especially in cortical regions associated with cognitive control and inhibitory functions. In a study by Bell et al. (2011), differences between cocaine groups abstinent for differing durations were observed bilaterally in the inferior longitudinal fasciculus, right anterior thalamic radiation, right ventral posterolateral nucleus of the thalamus, left superior corona radiata, superior longitudinal fasciculus bilaterally, right cingulum, and the white matter of the right pre-central gyrus. The findings suggested that certain specific white matter differences persist throughout abstinence, while other spatially distinctive differences discriminate as a function of abstinence duration. These differences may, therefore, represent brain changes that mark recovery from addiction. Similar findings have been found in heroin-dependent subjects from research in Liu's group. They found that fractional anisotropy was significantly decreased in specific brain regions of heroin-dependent patients ($p < 0.001$ uncorrected), including the frontal gyrus, the parietal lobule, the insula, and the corpus callosum. Thus, the presence of microstructural abnormality is found in the white matter of several brain regions of heroin-dependent patients (Yuan et al. 2009; Zhang et al. 2011).

Based on the current literature and our pilot findings presented herein, we are poised to further evaluate the effect of KB220Z on microstructural disruption of white matter in heroin addicts revealed by diffusion tensor imaging. Certainly, the coupling of multiple approaches (e.g., BOLD activation of dopaminergic pathways in the caudate-accumbens, attenuation of abnormal hyperactivity of the putamen in heroin-dependent subjects, and potential reduction in microstructural abnormalities in white matter by KB220Z) should ultimately lead to a novel safe DA agonist for prevention, tertiary treatment, and relapse attenuation in victims of Reward Deficiency Syndrome, especially carriers of reward gene polymorphisms.

Fearless moral inventory must include not only the drug of choice but also other Reward Deficiency Syndrome–related behaviors. This is so because the phenotype is not any particular drug or behavior of choice; it is indeed Reward Deficiency Syndrome. However, the inventory the individual is completing cannot be "right" or "wrong," because it his/her own list of resentments and evaluation of self. Moreover, the Big Book states, "No one among us has been able to maintain perfect adherence to any of these principles. The point is that we are willing to grow along spiritual lines. The principles we have set down are guides to progress. We claim spiritual progress rather than spiritual perfection." Several fourth steps may be taken by an individual over the course of his/her sobriety. Moreover, it is literally almost impossible for early recovering addicts to embrace Step 4 due to protracted abstinent impairments of brain reward circuitry, for example, in alcoholics, heroin addicts, and cocaine addicts. Unfortunately, this could be due to the chronic abuse of these powerful substances as an epigenetic phenomena, as well as possible inherited reward gene polymorphisms that occur at birth. It has been reasoned that one therapeutic target involves continued natural DA D2 activation as reflected in the preliminary fMRI research being conducted in China using KB220Z.

2.5 Step 5: Admitted to God, to Ourselves, and to Another Human Being the Exact Nature of Our Wrongs

In order to comprehend the meaning of the word "wrongs," it seems imperative to explore the nature of doing something wrong specifically as it relates to drug seeking. While this is very simplistic and of cause, the "wrongs" referred to by Step 5 involve very complex behaviors. We feel that it is important to explore human desires to reach "high pleasure states" through the use of psychoactive drugs and/or risk-taking behaviors, as well as the neurochemistry that supports the unleashing of behaviors that result in these wrongs.

It is important to understand that in AA parlance, "alcoholism" does not merely refer to one's addictive relationship with alcohol; it is a categorical condition of one's identity as an alcoholic, which includes a variety of different character defects. The "wrongs" this step is concerned with need not only refer to the misdeeds performed in the name of "drug-seeking" or under the influence of alcohol. It may be better to understand these wrongs as being performed in the name of *self-seeking,* or under the influence of alcoholism, a soul-sickness of a kind, for which, as the *Big Book* says, "our liquor was but a symptom" Thus, it is *all* wrongdoings that the individual is to list in this step, prior to reading it to another human being and God. More practically, however, these wrongs are found in the fourth step and read or "admitted" to another human being in the fifth step.

2.5.1 Understanding the High Mind

As of 2012, the United States has a total resident population of almost 312.8 million people (US Census Bureau). The total population of the United States at the turn of the twenty-first century was 281,421,906. The total number of people above the age of 12 years old was estimated at 249 million. More than 22 million Americans age 12 and older—nearly 9 % of the U.S. population—use illegal drugs, according to the government's 2010 National Survey on Drug Use and Health. We must then ask, who are the people who could, as Nancy Reagan suggested, "just say NO"? When almost half of the US population have indulged in illegal drug practices, when our presidential candidates are forced to dodge the tricky question of their history involving illegal drug use, and when almost every American has sloshed down a martini or two in their life time, there must be a reason, there must be a need, there must be a natural response for humans to imbibe at such high rates. There is even a more compelling question surrounding the millions who seek out high-risk novelty. Why do millions have this innate drive in face of putting themselves in Harms-way? Why are millions paying the price of their indiscretions in our jails, in hospitals, in wheel chairs and are lying dead in our cemeteries. What price must we pay for pleasure seeking or just plain getting "HIGH"? Maybe the answer lies within our brain. Maybe it is in our genome? Utilization of the candidate versus the common variant approach may be parsimonious as it relates to unraveling the addiction riddle. Here, we discuss evidence, theories, and conjecture about the "High Mind" and its relationship to evolutionary genetics and drug-seeking behavior as impacted by genetic polymorphisms. We consider the meaning of recent findings in genetic research including an exploration of the candidate versus the common variant approach to addiction, epigenetics, genetic memory, and the genotype–phenotype problem. We speculate about the neurological basis of pleasure seeking and addiction, the human condition, and the scope of societal judgments that effect multitudes in a global atmosphere where people are seeking "pleasure states." Is this natural tendency wrong or is it a consequence of our natural brain reward wiring?

A fundamental premise about the brain is that its workings—sometimes referred to as the "mind"—are the result of its anatomy, physiology, neurochemistry, and genome, and that is all. It is not our intent to argue or defend the mind/body dualism theories; rather, this is an attempt to understand the mind in terms of alterations in its "level of function."

It is interesting that as early human beings spread out from Africa starting around 60,000 years ago, they encountered environmental elements and challenges that they could not overcome with prehistoric technology. In this regard, many scientists falsely expected that analyses of our genome would indicate that an explosion of evolutionary mutations would have spread quickly throughout different populations. They theorized that these beneficial mutations would confer a greater survivability. However, Pritchard and Di Rienzo (2010) suggested that recent human evolution has occurred at a far slower pace than biologists had envisioned. While there have been examples of strong genetic mutations that have resulted in adaption to environmental pressure such as observed in Tibet, where transition to high altitudes resulted in gene mutation that favored this environmental shift (Moore et al. 2002). The genome actually contains few examples of very strong rapid natural selection. Instead most of the visible natural selection appears to have occurred over tens of thousands of years. It is noteworthy that the rate of change of most traits is very slow indeed and as such major adaptive shifts require stable conditions across millennia. Pritchard wrote (2010): "Thus 5,000 years from now the human milieu will no doubt be very different. But in the absence of large-scale genomic engineering, people themselves will probably be largely the same. With this in mind, then it may not be surprising to find a fairly high genetic influence on drug-/pleasure-seeking behavior in Homo *sapiens* possibly due to carrying a gene form that sets up the individual to be driven by a need to up-regulate specific proteins such as dopamine receptors to induce a feeling of well-being and happiness."

In (1972), Andrew Weil, in the best-selling book "*The Natural Mind,*" proclaimed that getting "high" is a time-honored tradition. Human beings are born with an innate need to get high, to experience periodically other states of consciousness, and the capacity for this experience is a capacity of the human nervous system. Often, external things, such as psychoactive drugs, like strong marijuana, seem to cause highs, but this is an illusion. In fact, it is due to the interaction of the drug and mesolimbic chemical messengers with the net effect of releasing DA at the reward site. Indeed, it is a pleasurable activity likened to a pre-orgasmic experience increasing ones libido. Weil and Rosen further characterized the "high" phenomena by correctly suggesting that some people may go from chocolates to morphine (Weil and Rosen 1983). Interestingly, the commonality between these two diverse substances resides in the content of chocolate that includes isoquinoline a substance shown to be opiate like.

In terms of science, the act of being high depends on one's receptor sensitivity to interact with certain chemical messengers including endorphins and DA. In reality, unless there are sufficient quanta of the so-called high molecules in the synapse, no high will be achieved. In essence, the genome of each individual may have a lot to do with one's sensitivity, need, and innate tolerance for the psychoactive effects of any substance and potential relapse (Blum et al. 2009a; Kirsch et al. 2006). Therefore, most people do not want to train their mind and slowly increase their ability to get high naturally. We must accept the fact that many people are going to rely on external things, including drugs and plant psychotropics, for their highs. We are also cognizant that craving behavior involves multiple neural pathways (see Fig. 2.4). It is important to be cognizant that the continual act of getting high with powerful psychoactive drugs could lead to addiction accompanied with many physiological changes both central and peripheral in nature (Gold 1993b). Furthermore, while it is well known that individuals self-medicate themselves to "feel normal," it is true that while this provides a pseudo feeling of happiness, ultimately it will lead to a down-regulation of DA receptors and as such ultimately will result in a feeling of unhappiness, depression, and more drug- or food-seeking behavior (Edge and Gold 2011; Blum et al. 2008, 2011b).

From the beginning, mankind has devoted considerable energy and ingenuity to "turning on" under such labels as "altering levels of consciousness" or as it used to be noted by Aldous Huxley, "opening the doors of perception." Through smoking, snorting, sniffing, eating, drinking, or mainlining, people of all cultures

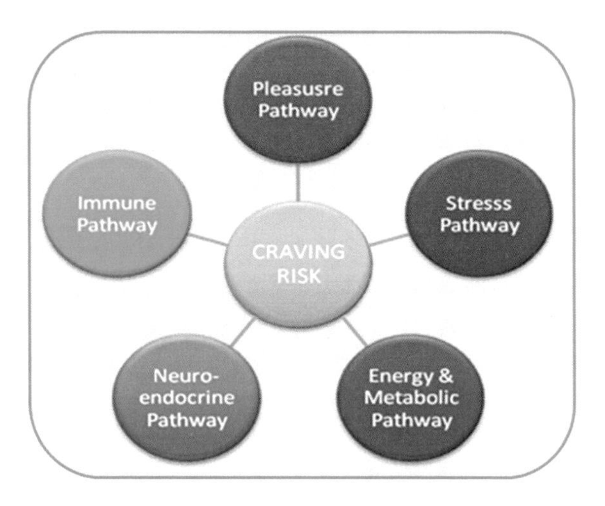

Fig. 2.4 Multiple neural pathways of craving. Craving behavior is due to multiple factors and involves a number of interacting pathways including pleasure, stress, immune, neuroendocrine, and energy and metabolic. These interconnecting pathways work in concert to induce a net release of dopamine in the NAc of the VTA in the mesolimbic system of the brain

have sought a little more than the standard view of reality. We could argue that today the problem of substance use disorder is symptomatic of more general societal problems, but we now know that certain gene forms can express a phenotype which actually predisposes an individual to wanting to get high compared to others who do not like to get high. In this regard, we know that individuals who carry the DRD2 A1 gene form love psychostimulants (cocaine) and those who carry the DRD2 A2 gene form hate psychostimulants (Persico et al. 1996). The continued understanding of neurogenetics and its role in drug-seeking behavior will be the subject of many experiments in the future, extending our existing knowledge of psychiatric genetics. The drug scene is nothing new and seems to be increasing in younger generation. It is the American way, a trillion dollar complex that pushes and pumps its produce into all facets of our life. Burger King cannot compare to the blazing RX sign in the sky. "If you have a pain, we have a pill." The principal questions at hand are "What quantity of drugs can we afford?" and "What is the system of barter?" Drugs or plant psychotropic abuse is not the primary issue, people are. It is not surprising that in the United States of America, two states (Washington and Colorado) have passed laws that legalize marijuana as a recreational substance. It may be difficult for some people to realize that people need other people as a basic human condition, but one of AA slogans is, "You alone can do it, but you can't do it alone." Human bonding is an adaptive evolutionary strategy that developed from a common neural circuitry underlying reproduction, parental/offspring attachment, and kinship-based groups (Dawkins 2009; Miller and Rogers 2001). Without interaction with another human, a person can become psychotic. Popular examples of the influence of human interaction on mental states are abundant. In the movie the "*Awakening,*" the Robert DeNiro character was suffering from a Parkinsonian-like tremor from the lack of brain DA, but when he danced with a loving caring female his tremors stopped. Howard Hughes, one of the wealthiest men in history, was a severely disturbed individual despite his riches. One can have all the riches in the world and all those things you think you really need to survive and prosper, but there is nothing like a beautiful smile, a handshake, or a hug or a kiss from another individual, or even an intelligent discussion to make you feel like a million bucks! People provide the real "highs." Barbara Streisand sang, "People who need people are the luckiest people in the world." However, in today's financially unstable (1.4 million bankruptcies 2009)—and frightening world, loneliness and alienation are commonplace. Where love, compassion and friendship are lacking, there is always chemistry to turn you on to synthetic "highs." While this might be true for many, there are also those who just like to get high through drugs and do it for no special reason except to feel good. It is for the party not for the escape. In whatever way happiness is sought, whether through other people, drugs, or sugar-coated placebos, the end result is that an individual strives in his own way to achieve happiness. However, this has resulted in a billion-dollar industry, that is, drug rehabilitation, with its 12,000 or more treatment centers.

While it is true that *Homo sapiens* in evolutionary terms are changing very slowly, it is also true that certain genetic traits such as genes that regulate pleasure seeking may be the exception. At this juncture, we do not know whether the DRD2 A1 allele is an older gene allele or is it newer than the DRD2 A2 allele. Understanding this will help explain the nature of the relationship, humans have with pleasure seeking and possibly how it benefits our survival. Certainly, carriers of the DRD2 A1 allele are more aggressive than carriers of the DRD2 A2 allele (Chen et al. 2005).

While it is true that neuroimaging studies of the brains of people addicted to drugs have helped to clarify the mechanisms of drug addiction, we must reflect on the question of how we address legally, the natural pursuit of pleasure seeking. Moreover, these studies and the initial work of Blum, Noble, and others (1991) have also helped to change the public's view of drug addiction from that of a moral violation or character flaw, to an understanding that pathological changes to brain structure make it very difficult for addicts to give up their addictions.

Abnormalities in prefrontal cortex of addicts create a feeling of need or craving that addicts know is irrational but cannot prevent. Prefrontal abnormalities also make it difficult to override compulsions to take drugs by exercising cognitive control. The main areas affected are the orbitofrontal cortex, which is important in maintaining attention to goals, and the anterior cingulate cortex, important in mediating the capacity to monitor and select action plans. Both areas receive stimulation from DA centers lower in the brain.

A steady influx of DA makes it difficult for addicts to shift their attention away from the goal of attaining drugs. It also fastens their attention to the motivational value of drugs, even though these drugs have long stopped providing pleasure. While the release of DA may result in ultimate pleasure states, its basic importance or relevance is that of sought-after goals. Addicts have a hard time turning their attention and their actions away from the goal of acquiring and consuming drugs. They are caught in a spiral of physical brain changes and the psychological consequences of those changes, leading to further changes.

Lastly, attempts to understand the "high mind" have eluded even the best neuroscientists. In approximately one-third of America, DA is a key genetically induced deficient neurotransmitter resulting in aberrant craving behavior and excessive pleasure seeking. Is it parsimonious that finding ways to enhance DA D2 density instead of blocking dopaminergic function may be the best strategy to unlock the elusive addiction riddle and attenuate abuse? Perhaps, the answer has been there all along! We ask: Are we dealing ourselves enough "*Dopamine for Dinner*?" ("What's for Dinner?" 2009).

2.5.2 Insights into the Mystery of Anxiety and Aggressive Behavior of the Alcoholic

One question that has been perplexing to the scientific community involves the basic neurobiology related to anxiety and to alcohol-induced violent outburst (a "wrong"). Certainly, anxiety is prevalent in the recovering alcoholic and it has been established that the condition is linked to the GABA receptor system. It has been further proposed that when one consumes alcohol, it is converted to acetaldehyde and this substance combines with DA to form a tetrahydroisoquinoline (TIQ) called beta-carboline (Blum et al. 1982c). This substance binds to benzodiazepine receptors and prevents the binding of the anti-anxiety substance GABA causing severe anxiety and aggression (Costa and Guidotti 1979). Whereas GABA reduces anxiety, beta-carboline induces it. Then, Costa and Guidotti (1983) found a new peptide in rat brains they called diazepam-binding inhibitor (DBI) that binds to benzodiazepine receptors and like beta-carboline induces anxiety (Guidotti et al. 1983). Some years later they found that DBI inhibits the binding of benzodiazepine to it receptors by competing for the sites. In fact, they also showed that when DBI is injected into rats, it causes anxiety as measured by the well-known Geller-Seifter conflict test. Most interesting they found that in males compared to females (except when females are menstruating), the DBI content of cerebrospinal fluid was significantly higher (see van Kammen et al. 1993). Based on this work, it has been concluded that the TIQ beta-carboline and DBI are natural anxiety agents in the brain, and that GABA is a natural anti-anxiety agent (Ferrero et al. 1986).

Further insights into the role of DBI in anxiety came when Guidotti collaborated with Alho and his group at the Research Laboratories of the Finnish State Alcohol Company. The following results stand out: (1) When alcohol-preferring rats were allowed to drink alcohol at will for a period of three months, the amount of DBI in the cerebellum and hypothalamus rose significantly, and (2) when non-alcohol-preferring rats were subjected to the same experiment, no such increase was found in DBI levels in the brain (Alho et al. 1987).

The results of this last experiment may provide at least a partial answer to the question of why the genetically predisposed individual or the chronic alcoholic so often develops intense anxiety (which is also tied to carrying the DRD2 A1 allele and polymorphisms of COMT; Blum 2007; Blum et al. 2012b), or aggressive/schizoid, or even violent behavior (a "wrong") after drinking (Blum et al. 1997). This violent behavior causing harm to others may simply be the effect of DBI, which has been shown to have powerful emotional side effects (Linnoila et al. 1983; Muhlenkamp et al. 1995).

For the interested reader, DBI has powerful multiple biological effects, as reviewed by Costa and Guidotti (1991). DBI is a 9-kD polypeptide that was first isolated in 1983 from rat brains by monitoring its ability to displace diazepam from the benzodiazepine recognition site located on the extracellular domain of the type A receptor for gamma-aminobutyric acid (GABAA receptor) and from the mitochondrial benzodiazepine receptor located on the outer mitochondrial membrane (Costa et al. 1975). In brain, DBI and its two major processing products [DBI 33–50, or octadecaneuropeptide, and DBI 17–50, or triakontatetraneuropeptide (TTN)] are unevenly distributed in neurons, with the highest concentrations of DBI (10–50 microMs) being present in the hypothalamus, amygdala, cerebellum, and discrete areas of the thalamus, hippocampus, and cortex. DBI is also present in specialized glial cells (astroglia and Bergmann glia) and in peripheral tissues. The neurobiological effects of DBI and its processing products in physiological and pathological conditions (hepatic encephalopathy, depression, panic) concentrations may, therefore, be explained by interactions with different types of benzodiazepine recognition sites. In addition, recent reports that DBI and some of its fragments inhibit (in nanomolar concentrations) glucose-induced insulin release from pancreatic islets and bind acyl-coenzyme A with high affinity, supports the hypothesis that DBI is a precursor of biologically active peptides with multiple actions in the brain and in peripheral tissues. Furthermore, it was found that the cerebrospinal fluid levels of DBI were higher in paranoid schizophrenics compared to chronic undifferentiated schizophrenic patients (Barbaccia et al. 1986).

Understanding our natural desire to obtain pleasure states and to admit "wrong doing" to God, ourselves, and those around us is no simple task and involves the consideration of not just our issue with "getting high" but rather with the toxic effects produced in the brain by continual exposure to these powerful substances. Their impact on brain reward networks is indeed physiological (e.g., increase in brain DBI). This can result in mental effects (anxiety and aggression) that also result in harmful behaviors with hurtful, harmful and sometimes fatal consequences not only to oneself but also to others.

2.6 Step 6: Were Entirely Ready to have God Remove All These Defects of Character

Step 6 is simply about willingness and readiness to change. Most of the discussion about moral character that follows is relevant to the exposition of Step 5. That being said, moral character or deficits of character refer to an evaluation of a particular individual's durable moral qualities. The concept of *character* can imply a variety of attributes including the existence or lack of virtues such as integrity, courage, fortitude, honesty, and loyalty, or of good behaviors or habits. Moral character primarily refers to the assemblage of qualities that distinguish one individual from another—although on a cultural level, the set of moral behaviors to which a social group adheres can be said to unite and define it culturally as distinct from others (e.g., Mormonism). Psychologist Lawrence Pervin (1960) defined moral character as "a disposition to express behavior in consistent patterns of functions across a range of situations."

The word "character" is derived from the ancient Greek word *charaktêr*, referring to a mark impressed upon a coin. Later, it came to mean a point by which one thing was told apart from others (Campbell and Bond 1982). The same Greek word is used in Christian Scripture (New Testament) as in most other Greek texts from this period. The word is defined the same way. There are two approaches when dealing with moral character: *Normative ethics*, which are moral standards that involve right and wrong conduct, a test of proper behavior by determining what is right and wrong; and *applied ethics,* which involve specific and controversial issues along with a moral choice and tend to be situations where people are either for or against the issue (Anscombe 1958). In 1982, Campbell and Bond proposed the following as major factors in influencing character and moral development: heredity, early childhood experience, modeling by important adults and older youth, peer influence, the general physical and social environment, the communications media, the teachings of schools and other institutions, and specific situations and roles that elicit corresponding behavior.

What is character? A dictionary describes character as the complex of mental and ethical traits marking a person. Character is who we really are and what we do: the accumulation of thoughts, values, words, and actions. People say you can achieve success by having good character (Perry and Körner 2011). But what really is good character? A person of good character thinks right and does right according to the core universal values that define the qualities of a good person: trustworthiness, respect, responsibility, fairness, caring, and citizenship. The Standard Encyclopedia of Philosophy provides a historical account of some important developments in philosophical approaches to moral character. A lot of attention is given to Plato, Aristotle, and Karl Marx's views. Marx accepts Aristotle's insight that virtue and good character are based on a sense of self-esteem and self-confidence. Plato believed that in order to have moral character, we must understand what contributes to our overall good and have our spirited and appetitive desires educated properly, so that they can agree with the guidance provided by the rational part of the soul. Abraham Lincoln once said, "Character is like a tree and reputation like its shadow. The shadow is what we think of it; the tree is the real thing" (*Lincoln's Own Stories*). According to the Bible, character is any behavior or activity that reflects the character of God. In general, Christians believe that this means that the morally correct thing to do is reflect the character of the creator. Christian character also is defined as exhibiting the "Fruits of the Spirit" as defined in the Bible, specifically in Galatians 5:22–23: "But the fruit of the Spirit is love, joy, peace, patience, kindness, goodness, faithfulness, gentleness and self-control. Against such things there is no law." Interestingly, there is controversy about the true meaning of character. A moral character trait (genetic) is a character trait for which the agent is morally responsible. If moral responsibility is impossible, however, then agents cannot be held responsible for their character traits or for the behaviors that result from those character traits.

A similar argument also has been advocated by Bruce Waller. According to Waller, no one is "morally responsible for her character or deliberative powers or for the results that flow from them." Given the fact that she/he was shaped to have such characteristics by environment (or evolutionary-genetic) forces far beyond a person's control, she/he deserves no blame nor praise (Doris 2000; Homiak 2008; Huitt 2004; Lawrence 1994).

The character defects addressed by the twelve-step program are those negative qualities and tendencies that are possessed by each and every person, but undoubtedly alcoholism and drug dependence make them worse. These moral deficits are the causes of the addictive behavior, and it is the compulsion of addiction that overcomes the will and the conscience and leads addicts to make ethical and moral compromises. In other words, there is a dynamic, reciprocal relationship between these elements.

Character defects are reflected in exploitative behaviors that suggest personal qualities such as jealously, self-pity, greed, grandiosity, meanness, selfishness, arrogance, and callousness. Consequently, alcoholics and addicts often commit errors in judgment that hurt others. AA, NA, and other self-help groups have long recognized this unfortunate reality. Alcoholics need to be honest about these wrongs and errors in order to be able to experience a spiritual renewal.

2.6.1 Honesty

To achieve Step 6 and have a readiness for God to remove character defects, especially as it relates to marriages or friendships, problems with children, business relationships between other ways in which he/she has hurt other people, either willfully or accidentally needs honest admission. As stated in the *Big Book*, "Those who do not recover are people who cannot or will not completely give themselves to this simple program, usually men and women who are constitutionally incapable of being honest with themselves." Even the concept of honesty may be rooted in an individual's genetic makeup. In the world of molecular neurobiology, lying or being dishonest can be measured by using the Defense Style Questionnaire (Comings et al. 1995).

The Defense Style Questionnaire was originally developed by Bond and colleagues (Bond et al. 1983; Bond 1986; Andrews et al. 1989; Frank et al. 2006) to evaluate a subject's style of dealing with conflict. The sub-scales are divided into mature defenses (sublimation, humor, anticipation, and suppression); neurotic defenses (undoing, pseudo-altruism, idealization, and reaction formation); and immature defenses (projection, passive aggression, acting out, isolation, devaluation, autistic fantasy, denial, displacement, dissociation, splitting, rationalization, and somatization). Patients with psychoses or anxiety disorders (Bond and Vaillant 1986; Vaillant et al. 1986; Pollock and Andrews 1989), a history of abusing their children (Brennan et al. 1990), or of smoking or drinking during pregnancy (Kesby et al. 1991) scored higher on immature defenses than normal controls. In a large, prospective study of adult twins, Andrews (1991) determined that 38 % of the variance in the Defense Style Questionnaire could be attributed to genetic factors, half of which seemed specific to the defense style and half to genetic factors.

The Defense Style Questionnaire was administered to Caucasian males consisting of 123 subjects from a V.A. addiction treatment unit, 42 Tourette's syndrome subjects, and 49 controls. For the addiction treatment unit and Tourette's subjects, there was a significant decrease in the mean score for mature defenses and a significant increase in mean score for immature defenses compared to controls. Many of the individual sub-scores showed the same significant differences. DA D2 receptor (DRD2) gene haplotypes, identified by allele-specific polymerase chain reaction of two mutations (G/T and C/T) 241 base pairs apart, were determined in 57 of the addiction treatment unit subjects and 42 of the controls. Subjects with the 1 haplotype tended to show a decrease in mature and an increase in neurotic and immature defense styles compared to those without the 1 haplotype. Of the eight times that the sub-scale scores were significant for haplotype 1 versus non-1, they were always in this direction. These results suggest that the DRD2 locus is one factor controlling defense styles or "lying." The difference in the mean scores between controls and substance abuse subjects indicated that other genes and environmental factors also play a role.

Although it is possible to define character in a moralistic sense, it is very difficult to assign responsibility for defects of character and the bad decisions and consequence since character is shaped by genetic (evolutionary) forces far beyond a person's control. With this stated, it is argued that environmental elements especially in childhood may also require rethinking in terms of blame and or even praise of an individual act. This idea supports the idea in the sixth step that the removal of character defects is the province of a higher power. Clinicians should be cognizant that for the individual, achievement of this step requires deep character analysis, painful realization, and ability to dissociate oneself (present) from the past self. It should also be noted that carriers of the DRD2 gene polymorphism (risk for addiction) will have great difficulty in achieving honesty.

2.7 Step 7: Humbly Asked Him to Remove Our Shortcomings

The shortcomings in Step 7 refer to the character defects addressed in Step 6, which were revealed through the inventory in Step 4. The AA co-founders were at pains to point out that the individual was not excused for his/her wrongdoings, despite having emerged from the illness of alcoholism. Hence, the Steps describe a process of taking ownership and responsibility for one's actions, both past and present, as a means of obtaining the "spiritual awakening" described later in Step 12.

The definition of "**shortcomings**" is a failure, defect, or deficiency in conduct, condition, thought, ability, etc.: a social shortcoming; a shortcoming of one's philosophy. It is a character flaw or weakness, and in times before there was a disease concept of alcoholism, excessive drinking was considered to be a weakness of willpower (Dean and Poremba 1983). In fact, prior to Blum's research along with Ernest Noble (Blum et al. 1990) in associating the first gene polymorphism for severe alcoholism, less than half of the United States believed that alcoholism was a reflection of individuals' shortcomings, flawed personality, and weaknesses. Interestingly, the day after this finding was publicized, a Gallup Poll indicated that for the first time, the acceptance of alcoholism as a genetic disorder occurred in 56 % of those individuals polled.

It is important to consider the fact that these shortcomings or deficiencies arise from the reward circuitry of our brains. Dopaminergic neurons in humans play a major role in a wide range of behaviors, including impulsivity, aggression, sexual behavior, appetite, reward, and regulation of pituitary hormones (Iversen and Alpert 1982; Louilot et al. 1989; Smith and Schneider 1988; Blackburn et al. 1992; Lorenzi et al. 1980; Routtenberg 1987; Wise and Rompre 1989). The isolation of the DA D2 receptor (DRD2) gene in rats and humans by Civelli and co-workers (Bunzow et al. 1988; Grandy et al. 1989) opened the way for studies of the potential role of genetic variants of this locus in human behavior. Grandy et al. (1989) described a *Taq1* polymorphism located 3' to the gene. This was termed the *Taq1*A allele. In 1990, Blum et al. reported that the presence of A1 allele of the DA D2 receptor gene correctly classified 77 % of alcoholics, and its absence classified 72 % of non-alcoholics. Since these studies were performed on brain tissue of controls and alcoholics, they were able to directly measure the DA binding in the striatum of their subjects (Noble et al. 1991). This showed that both controls and alcoholics carrying the D2A1 allele had a significantly lower 3H-spiperone B than those with the D2A2 allele. These results suggested that the *Taq* AI/A2 polymorphism was in linkage disequilibrium (Hague et al. 1991; Uhl et al. 1993) with mutations affecting the expression of the DRD2 gene.

Subsequent studies in alcoholics have both confirmed and denied a relationship between the DRD2 A1 allele and alcoholism. A review of all the data to date (3386 articles in PUBMED 1/3/13) suggests that alcoholism is a heterogeneous disorder, and some forms are associated with the D2A1 allele (Noble 1993). In addition to the effect of the D2A1 allele on alcoholism, many reports also concur that the DRD2 gene is correlated with severity of drug addiction (Noble et al. 1993; Smith et al. 1992; Comings et al. 1993, 1994; O'Hara et al. 1993; Nyman et al. 2012). The frequency of the D2A1 allele was also significantly increased in individuals with attention deficit hyperactivity disorder, Tourette's syndrome, conduct disorder, and post-traumatic stress disorder (Comings et al. 1991; Comings 1994a, b). There are many reports showing a significant correlation between the D2AI allele and pathological gambling (Comings et al. 1996a) and smoking (Comings et al. 1996b).

Seeking relief from known shortcomings and attempting to have enough faith, so that these character flaws can be removed by a higher power, could be very challenging to the recovering individual who will have to face the fact that honest effort will not always succeed in getting us what we truly want from life and could result in severe depression. It has been shown that depression can reflect a feeling of hopelessness for anyone with a chronic illness including addiction (Myung et al. 2010; McLellan 2000).

Polymorphisms of the 5-HT transporter gene are involved in hypodopaminergic function. The 5-HT transporter gene is also one of the most studied genes for depression/hopelessness. Myung et al. (2010) examined the association of 5-HT transporter gene polymorphisms with chronicity and recurrent tendency of depression in Korean subjects. This cross-sectional study involved 252 patients with major depression. Patients were genotyped for s/l polymorphisms in 5-HTT promoter region (5-HTTLPR), s/l variation in second intron of the 5-HTT gene (5-HTT VNTR intron2). Chronicity was associated with 5-HTTLPR. Patients with l/l had higher rate of chronicity than the other patients (l/l vs. s/l or s/s; odds ratio, 4.45; 95 % confidence interval, 1.59–12.46; $p = 0.005$; logistic regression analysis). Recurrent tendency was not associated with 5-HTTLPR. Chronicity and recurrent tendency were not associated with 5-HTT VNTR intron 2. These results suggested that chronic depression is associated with 5-HTTLPR and as such important to test for in subjects with addictive risk.

In addition, work by Wojnar et al. (2009) examined relationships between genetic markers of central 5-HT and DA function, and risk for post-treatment relapse in a sample of alcohol-dependent patients. Of 154 eligible patients, 123 (80 %) completed follow-up and 48 % ($n = 59$) of these individuals relapsed. Patients with the Val allele in the Val66Met brain-derived neurotrophic factor polymorphism and the Met allele in the Val158Met COMT polymorphism were more likely to relapse. Only the brain-derived neurotrophic factor Val/Val genotype predicted post-treatment relapse [odds ratio (OR) = 2.62; $p = 0.019$] and time to relapse (OR = 2.57; $p = 0.002$), after adjusting for baseline measures and other significant genetic markers. When the analysis was restricted to patients with a family history of alcohol dependence ($n = 73$), the associations between the brain-derived neurotrophic factor Val/Val genotype and relapse (OR = 5.76, $p = 0.0045$) and time to relapse (hazard ratio = 4.93, $p = 0.001$) were even stronger. The Val66Met brain-derived neurotrophic factor gene polymorphism (a gene involved in the regulation of brain DA function) was associated with a higher risk and earlier occurrence of relapse among patients treated for alcohol dependence. The study suggests a relationship between genetic markers and treatment outcomes in alcohol dependence and underscores the importance of genes and shortcomings such as hopelessness and other mood issues.

Being humble must be accompanied with both gratitude and grace. The concept "turning it over" and "letting go" and let God remove our shortcomings is not easily accomplished. To be humble is akin to having gratitude for the things we have and the idea of moving forward. Statements of spiritual faith and being humble challenge the recovering person to face the fact that good intentions and honest effort alone will not always succeed in getting him or her what is truly wanted from life. In turn, and supported by genetic predisposition, this could lead to chronic depression and relapse. However, it is noteworthy that the twelve-step program and the traditions together ask the person to believe that evil and brutishness, injustice and cruelty will not necessarily win out in end. Being humble and having faith indeed paves the way to advocate neither passivity nor hopelessness; on the contrary, they express the belief that our shortcomings can be removed by our willingness to believe that things can work out for the best in the long run. Agreeing to let God remove our shortcomings helps the individual to recognize and accept that while we are responsible for making an effort to get what we need, we cannot control the final outcome of that effort.

2.8 Step 8: Made a List of All Persons We had Harmed, and Became Willing to Make Amends to Them All

> Almost without exception, alcoholics are tortured by loneliness. Even before our drinking (or drugging) got bad and people began to cut us off, nearly all of us suffered the feeling that we didn't quite belong (*As Bill Sees It*).

Much of the data for Step 8 comes from the inventory conducted in Step 4. The eighth step is simply about willingness to change. In Step 8, alcoholics and addicts who find the courage to face their moral mistakes can humbly seek forgiveness from a higher power, and in some cases do something to make up for them. Accordingly, patients who admit to their defects will be able to make amends. Making amends appropriately can heal guilt and shame. Getting to this point does not occur easily and is best approached after a sustained period of sobriety. Thus, much of the information considered in this section also is suited for the next section, where "direct amends" actually are being made.

2.8.1 Friendship

There is a new science related to friendships, involving molecular genetics, especially as it relates to happiness and neurochemical interactions. Fowler et al. (2009) suggest that social networks exhibit strikingly systematic patterns across a wide range of human contexts. Although genetic variation accounts for a significant portion of the variation in many complex social behaviors, the heritability of egocentric social network attributes is unknown. However, they show that three of these attributes (in-degree, transitivity, and centrality) are heritable. These results suggest that natural selection may have played a role in the evolution of social networks. They also suggest that modeling intrinsic variation in network attributes may be important for understanding the way genes affect human behaviors and the way these behaviors spread from person to person (Fowler et al. 2009). In the same year, Jackson (2009) reviewed Fowlers' work and commented on the importance of these new findings. He asked, "Who becomes the most central individual in a society and why? What determines how many friends a given individual has? What determines how clustered or tightly knit the friendships in a society are?" According to Jackson, "in a set of important and original new findings reported in PNAS, Proceedings of the National Academy of Science (PNAS), Fowler, Dawes, and Christakis provide evidence that network characteristics such as those mentioned above are heritable: that is, they show that an increase in the overlap in genetic material in twins corresponds to an increase in the covariation of some of their social network characteristics."

The heritability of network characteristics is important because of its implications for how networks form. Given that social networks play important roles in determining a wide variety of things ranging from employment and wages to the spread of disease (Kendler and Baker 2007; Iervolino et al. 2002; Guo 2006), it is important to understand why networks exhibit the patterns that they do. Although it is well established that personal characteristics and behaviors play critical roles in determining who interacts with whom (Fowler et al. 2009; Dudley and File 2007), Fowler et al.'s analysis (Burt 2009) suggests that genetic traits may influence individuals in terms of their social behavior, for instance, by having genetic predispositions regarding characteristics such as the tendency of an individual to introduce his or her friends to each other.

Interestingly, according to Dick et al. (2007), gender and gender of friends moderate the associations between friends' behavior and adolescents' alcohol use, with evidence that girls, and those with opposite-sex friends, may be more susceptible to friends' influence. Genetically informative analyses suggest that similarity in alcohol use between adolescents and their friends is mediated, at least partially, through environmental pathways (Dick et al. 2007).

As we stated earlier, Fowler and associates (Fowler et al. 2011) pointed out that humans tend to associate with other humans who have similar characteristics. Moreover, humans are unusual as a species in that virtually all individuals form stable, non-reproductive unions to one or more friends. Along these lines, a 2002 study conducted at the University of Illinois by Diener and Seligman found that the most salient characteristics shared by the 10 % of students with the highest levels of happiness and the fewest signs of depression were their strong ties to friends and family and commitment to spending time with them. "Word needs to be spread," concludes Diener. "It is important to work on social skills, close interpersonal ties and social support in order to be happy" (Diener et al. 2006).

2.8.2 Families

Our laboratory (Blum et al. 2009a, 2011a) has found evidence that family members exhibiting multiple Reward Deficiency Syndrome types of behaviors (i.e., drug and alcohol addiction, smoking, sex addiction, pathological gambling, violence behavior (Haller et al. 1996), juvenile delinquency, criminal behavior, ADHD, etc.) marry other individuals possessing the A1 allele of the DRD2 gene 100 % of the time in one family. This lends further support to the old folk concept, *Birds of a Feather Flock Together*. It has been observed in addiction rehabilitation clinics and psychiatric hospitals that many individuals become friends with other drug-addicted persons (Blum et al. 2013).

2.8.3 Happiness

Of course, happiness is not a static state. Even the happiest of people—the cheeriest 10 %—feel blue at times. And even the "bluest," still have their moments of joy (Bruijnzeel et al. 2004). That has presented a challenge to social scientists trying to measure happiness and friendship. The simple fact that happiness is inherently subjective makes the challenge even more difficult. In a questionnaire, detailing everything they did on the previous day, and whom they were with at the time, 900 women in Texas rated a range of feelings during each episode (happy, impatient, depressed, worried, tired, etc.). Results were surprising. It turned out that the five most pleasurable and rewarding (positive) activities for these women were (in descending order) sex, socializing, relaxing, praying or meditating, and eating. Exercising and watching TV were not far behind. But way down the list was "taking care of my children," which ranked below cooking and only slightly above housework. The results of this rating should not be interpreted as meaning these activities were viewed as the most beneficial, productive, or important—just activities associated with greatest feelings of pleasure. Overall happiness is not merely the sum of our happy moments minus the sum of our angry or sad ones. In other work, Kahneman (2006) proposed the belief that high income is associated with good mood is widespread but mostly illusory. People with above-average income are relatively satisfied with their lives but are barely happier than others in moment-to-moment experience, tend to be tenser, and do not spend more time in particularly enjoyable activities. Moreover, the effect of income on life satisfaction seems to be transient. It has been argued that people exaggerate the contribution of income to happiness because they focus, in part, on conventional achievements when evaluating their life or the lives of others (Kahneman et al. 2006).

Guo et al. (2007) looked for a link between social behavior (morality) and the DA D2 receptor gene by assessing delinquency rates in teenagers. The study was based on a cohort of more than 2,500 adolescents and young adults in the National Longitudinal Study of Adolescent Health in the United States. For DRD2, the trajectory of serious delinquency for the heterozygotes (A1/A2) is about 20 % higher than the A2/A2 genotype and about twice as high as the A1/A1 genotype, a phenomenon sometimes described as heterosis (Comings and MacMurray 2000) (LR test, $p = 0.005$). The findings on violent delinquency closely resembled those on serious delinquency and depression (Haeffel et al. 2008).

In humans, dominance has been linked to heritable personality traits (Mehrabian and Blum 1996); furthermore, superior status interacts with multiple neurotransmitters such as DA D2/D3 receptor binding, whereas high binding associates with higher social status (Martinez et al. 2010) and neuroendocrine (Sapolsky 2005) systems and can be automatically and efficiently inferred (Moors and De Houwer 2005). This indicates the existence of biological systems that process social rank or social hierarchy information. An fMRI study by Zink et al. (2008) provides a characterization of the neural correlates associated with processing social hierarchies in humans. The study demonstrated that brain responses to superiority and inferiority are dissociable, even in the absence of explicit competition, both when encountering an individual of a particular status and when faced with an outcome that can affect one's current position in the hierarchy. The researchers found that viewing a superior individual differentially engaged perceptual-attentional, saliency, and cognitive systems, notably dorsolateral prefrontal cortex. Furthermore, social hierarchical consequences of performance were neurally dissociable and of comparable salience to monetary reward, providing a neural basis for the high motivational value of status. This work underscores the importance of hierarchy status in social networks linking status to brain reward circuitry, a site of emotion and well-being.

In fact, according to Fowler and Christakis (2008), people's happiness depends on the happiness of others with whom they are connected. They found that clusters of happy and unhappy people are visible in the network, and the relationship between people's happiness extends up to three degrees of separation (for example, to the friends of one's friends' friends). People who are surrounded by many happy people and those who are central in the network are more likely to become happy in the future. Specifically, a friend who lives within a mile (about 1.6 km) and who becomes happy increases the probability that a person is happy by 25 % (95 % confidence interval 1 to 57 %). Similar effects are seen in co-resident spouses (8, 0.2 to 16 %), siblings who live within a mile (14, 1 to 28 %), and next-door neighbors (34, 7 to 70 %).

Using this development of a research model to define social networks (Fowler et al. 2009), Fowler et al. (2011) correlated genotypes in friendship networks. Employing available genotype data derived from the both the National Longitudinal Study of Adolescent Health and the Framingham Heart Study, they found that DRD2 A1 allele is positively correlated (homophily), and CYP2A6 (SNP rs1801272) is negatively correlated (heterophily). These unique results showed that homophily and heterophily occur on an allelic level. The results suggested that association tests should include friends' genes and that theories of evolution should take into account the fact that humans might, in some sense, be metagenomic with respect to the humans around them (again supporting the concept that *Birds of a Feather Flock Together*; see Fig. 2.5).

Germaine to the subject related to the role of reward genes and social networks, it is important to point out the original study by Blum et al. (1990) associating the DRD2 A1 allele and sever alcoholism. Further, the DRD2 A1 allele has been associated with a number of Reward Deficiency Syndrome behaviors (Chen et al.

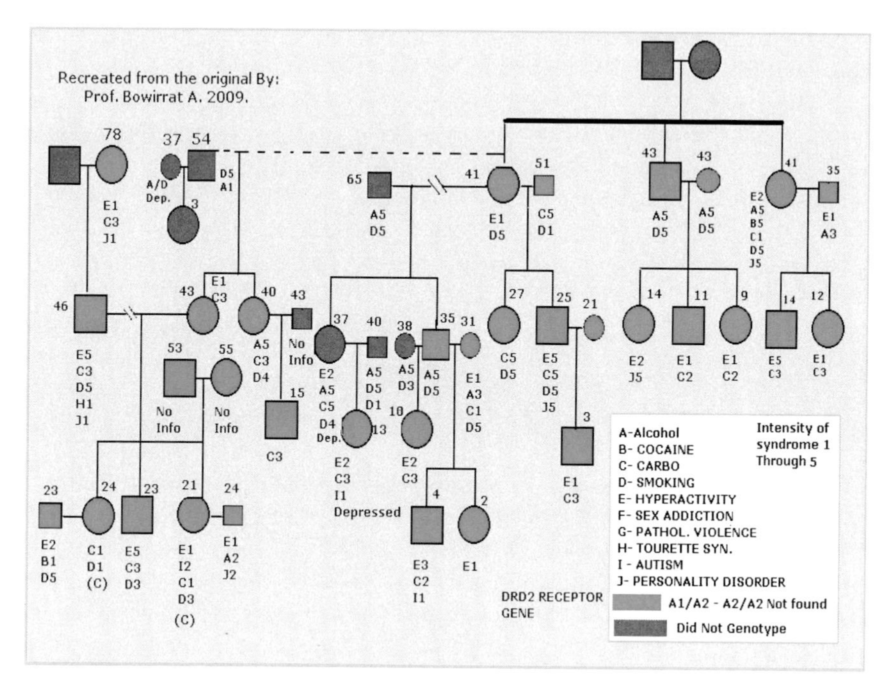

Fig. 2.5 Generational genotyping for dopaminergic genes in a family showing RDS behaviors. Genotype results of the dopamine D2 receptor gene (DRD2) polymorphism of family A ($n = 32$) identified with multiple Reward Deficiency Syndrome (RDS) behaviors; 100 % of probands married a mate carrying the DRD2 A1 allele (Blum et al. 2011a modified)

2011) including body mass index (Noble et al. 1994; Comings et al. 1993, 1996; Wang et al. 2001, 2011a; Blum et al. 1996a).

Coupling this understanding it is not surprising that Rosenquist et al. (2010) remarkably found that clusters of drinkers and abstainers were present in the network at all time points, and the clusters extended to three degrees of separation. These clusters were not only due to selective formation of social ties among drinkers, but they also seemed to reflect interpersonal influence. Accordingly, changes in the alcohol consumption behavior of a person's social network had a statistically significant effect on that person's subsequent alcohol consumption behavior. The behaviors of immediate neighbors and coworkers were not significantly associated with a person's drinking behavior, but the behavior of relatives and friends was associated. Similarly, Christakis and Fowler (2007) found clusters of obese persons with a body mass index greater than or equal to 30, who were present in the network at all time points, and the clusters extended to three degrees of separation. Accordingly, these clusters did not appear to be solely attributable to the selective formation of social ties among obese persons. A person's chances of becoming obese increased by 57 % (95 % confidence interval [CI], 6–123) if he or she had a friend who became obese in a given interval. Among pairs of adult siblings, if one sibling became obese, the chance that the other would become obese increased by 40 % (95 % CI, 21–60). Further, if one spouse became obese, the likelihood that the other spouse would become obese increased by 37 % (95 % CI, 7–73). These effects were not seen among neighbors in the immediate geographic location. Persons of the same sex had relatively greater influence on each other than those of the opposite sex.

To reiterate, these findings are in agreement with our own findings whereby genotyping for at least the DRD2 A1 allele using Reward Deficiency Syndrome as a generalized phenotype in a five generational genotype study showed that 100 % of family members carrying the DRD2 A1 allele married a person who also carried the same genotype (Blum et al. 2008; Blum et al. 2011a, b) (Fig. 2.5).

It is not easy to make amends especially to people who are not only our friends but people whom we love. Step 8 does not come early in one's sobriety but only after periods of being clean and sober. However, once an individual accomplishes this arduous task, he or she will be able to move forward in the path of recovery. In terms of connecting the dots, it is important for clinicians to realize that the old adage of "Birds of a feather flock together" may be inheritable by virtue of friends seeking friends who not only have similar characteristic (maybe even drinking, drugging, and eating), but also similar genotypes, such as the DRD2 A1 allele. So that when the alcoholic, for example, is asked to make amends and also eliminate certain friends that would not be conducive to their recovery, we need to be cognizant about going against the genetic grain. Thus, on a molecular neurobiological level, it is easily said but not easily done. A form of happiness is that people live in social networks that are comfortable. Making amends for the hurt may not re-establish trust but may help assuage guilt and shame. Here, it may be helpful to consider the genetic predisposition of families to Reward Deficiency Syndrome behaviors.

2.9 Step 9: Made Direct Amends to Such People Wherever Possible, Except When to do so Would Injure Them or Others

Addiction is a family disease with a high genetic heritability, and the addict might want to develop genograms to help trace family generations of addiction. These genograms could be used to examine family patterns in marriage (consider the DRD2A1 study whereby people married people with the same polymorphisms) and career choices, for example, or in lifestyle, or in parenting.

Moreover, high levels of parental rule-setting are associated with lower levels of adolescent alcohol use and delay of initiation of drinking. Van der Zwaluw et al. (2010) tested whether DRD2 TaqI A (rs1800497) genotype interacts with alcohol-specific parenting practices in predicting the uptake of regular adolescent alcohol use. DRD2 genotype interacted with parental rule-setting on adolescent alcohol use over time: adolescents, with parents highly permissive toward alcohol consumption and carrying a genotype with the DRD2 A1 (rs1800497T) allele, used significantly more alcohol over time than adolescents without these characteristics. The DRD2 genotype may pose an increased risk for alcohol use and abuse, depending on the presence of environmental risk factors, such as alcohol-specific parenting (van der Zwaluw et al. 2010).

Furthermore, in family therapy, genograms have been used to help patients gain a better understanding of the ways family dysfunction is perpetuated, including the patients choosing to break family patterns and build a different and hopefully healthier lifestyle than other family members have done in the past. A second purpose of the genogram is to highlight the concept of alcoholism and drug addiction as "family illnesses" (i.e., Reward Deficiency Syndrome). Finally, the genogram is intended to serve as a catalyst to explain how alcoholism and addiction (all types) have harmed not only the patient but others in his/her family, including previous generations.

While we are cognizant of free will, we must not be so naive that we underestimate the relationship between our basic social behaviors including parenting and biology. The study of genes potentially promises a better understanding of the constraints imposed on basic behavior (see Aristotle, 1996). Thus, we agree with Fowler and associates, and we argue further that biologists and social scientists must work together to advance a new science of human nature (Carioppo and Patrick 2008).

In simple terms, can we as scientists reduce the state of happiness and love of others to molecular rearrangements leading to gene polymorphisms? If indeed this were a simple matter, then why not consider the following: Love relationships relate to polymorphisms of the DRD2 gene with carriers of A1 alleles forming short term relationships involving erotic love styles. However, carriers with the appropriate 5-HT polymorphism would be potentially happier because they could form lasting relationships having romantic love styles. If that isn't enough, consider the fact that DRD2 A2 carriers are more likely to have social attachments compared to A1 carriers. This is further supported by earlier work from our laboratory showing the significant association of schizoid/avoidant behaviors in A1 carriers compared to A2 carriers (Blum et al. 1997). It is well established that schizoid/avoidant behavior occurs in people who are less passionate and cannot form meaningful relationships or attachment. Couple this with genospirituality and the probability, albeit small, of genospirituality engineering, and other as yet unidentified gene polymorphisms, and what emerges is a complex map of human nature tied to the unconscious state of happiness. There are multiple genes involved in the state of happiness that are interactive and thus affect reward type of behaviors (Charlton 2008) and of cause our need for friendships in spite of our shortcomings that induced "harm" to others (see Fig. 2.6).

More importantly, we must consider whether or not making amends to people we care about (or even some whom we do not) will hurt them in the long run. It would be reasonable to make amends considering the above because we as human beings need attachment to others.

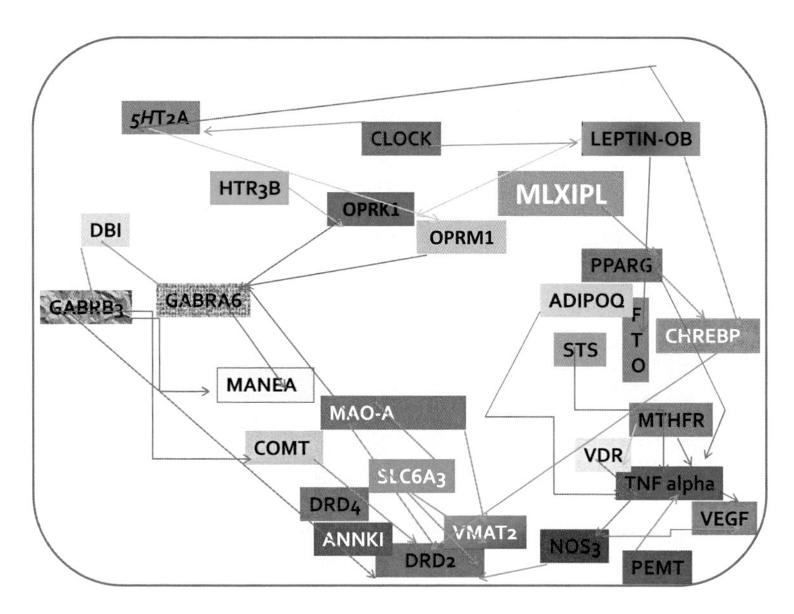

Fig. 2.6 Happiness gene map. The interaction of multiple genes involved in the net release of dopamine in the reward system VTA-NAc, ultimately leading to happiness (Blum et al. 2009a)

Based on a number of social science studies, it is well established that the behavioral characteristic known as attachment is tied to happiness (Carioppo and Patrick 2008). In their book, "*Loneliness: Human Nature and the Need for Social Connection*" (2008), modern day philosophers John T. Carioppo and William Patrick suggested that isolation can be harmful to your health, just as is smoking or a sedentary lifestyle. A large part of this effect is driven by the subjective sense of social isolation we call loneliness. Humans are far more intertwined, hardwired, and interdependent physiologically as well as psychologically than our cultural prejudices have allowed us to acknowledge. There is an African proverb that states: "If you want to go fast, go alone; if you want to go far, go together."

In terms of wellness, little is known about the genes that may regulate personality traits involved in the overall phenotype "well-being." Weiss et al. (2008) used a representative sample of 973 twin pairs to test the hypothesis that heritable differences in subjective well-being are entirely accounted for by the genetic architecture of the Five-Factor Model's personality domains. Their results supported this model. Specifically, subjective well-being was accounted for by unique genetic influences from Neuroticism, Extraversion, and Conscientiousness, and by a common genetic factor that influenced all five personality domains in the directions of low Neuroticism and high Extraversion, Openness, Agreeableness, and Conscientiousness. These findings indicate that subjective well-being is linked to personality by common genes. Personality may form an *affective reserve* relevant to set point maintenance and changes in set point over time. Other results also support a differentiated view that implies that genes and environment

(Pervin 1994) together play important roles in the associations between well-being and health (Røysamb et al. 2003*)*.

Finally, it is very interesting that in Bhutan and other ancient Eastern cultures, people believe that enlightenment through multiple paths, including meditation, yoga, and Buddhist spiritual teachings, can lead the way to satisfaction and fulfillment. In 1972, the Bhutanese king proclaimed that instead of measuring success by wealth or the Gross National Product, it should be measured by Gross National Happiness. Through many incarnations, one may become enlightened and reach the ultimate state of nirvana. Buddha described nirvana as the perfect peace of the state of mind that is free from craving, anger, and other afflictive states (*kilesa*). The subject is at peace with the world, has compassion for all, and gives up obsessions and fixations. This peace is achieved when the existing volitional formations are pacified, and the conditions for the production of new ones are eradicated. In nirvana, the root causes of cravings and aversions have been extinguished, such that one is no longer subject to human suffering (*dukkha*) or further states of rebirth in (*samsara)*.

With this in mind and being grateful for having this uplifting cultural mandate, one could make a suggestion that NIRVANA is indeed an important acronym whereby it could be defined as *Neurotransmitter Interaction at Reward Ventral Tegmental Accumbens Leading to Neuronal Adaptation*, or Happiness (see Fig. 2.7).

It is not easy to achieve happiness and peace especially when the alcoholic or addict is faced with taking responsibility for hurting others with whom he or she has relationships while drinking and drugging. An obvious source of injury to relationships caused by addiction is the "abandonment" of a spouse or significant

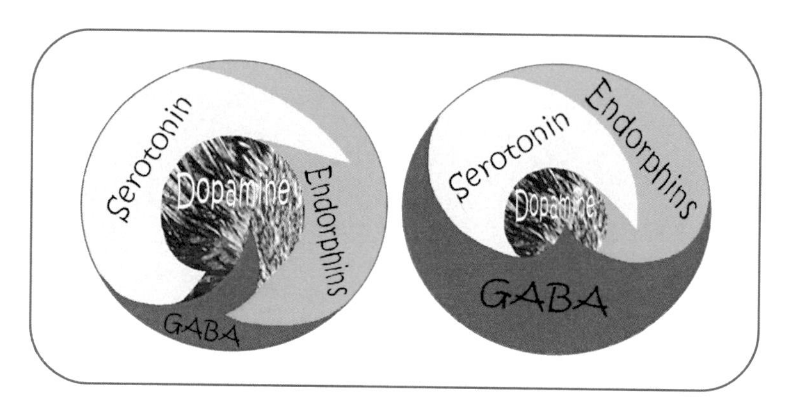

Fig. 2.7 Happy versus Unhappy Brain. *Happy Brain* (*left*) represents the normal physiological state of the neurotransmitter interaction at the mesolimbic region of the brain. Briefly, serotonin in the hypothalamus stimulates neuronal projections of methionine-enkephalin in the hypothalamus that, in turn, inhibits the release of GABA in the substantia nigra, thereby allowing for the normal amount of dopamine to be released at the Nucleus Accumbens (NAc); reward site of the brain. *Unhappy Brain* (*right*) represents hypodopaminergic function of the mesolimbic region of the brain. The hypodopaminergic state is due to gene polymorphisms as well as environmental elements, including both stress and neurotoxicity from aberrant abuse of psychoactive drugs (i.e., alcohol, heroin, cocaine, etc.) and genetic variables [Blum et al. 2012a with permission]

other for alcohol and/or drugs. Victims of Reward Deficiency Syndrome must take responsibility for this abandonment of loved ones. Furthermore, addicts may have been very abusive (both physically and emotionally) during their active addiction. Before any amends can be made, the addict is asked in Step 8 to take an inventory of all persons harmed, which can easily evoke intense feelings of guilt and shame. It also requires overcoming denial and being willing to make amends. In Step 9, the achievement of making amends (except where doing so would cause no further injury) is subject to correlations between genes, friendships, and relationships. As noted in the research summarized above, relationships and happiness are based on neuronal hard wiring, and this presents both a formidable challenge and clarity as to how to achieve effective healing in recovery. The degree to which the person can make amends to others (without harm or hurt) is tantamount to a healthy recovery and importantly the attainment of happiness. This can be facilitated through the positive natural release of DA in reward centers of the brain.

2.10 Step 10: Continued to Take Personal Inventory and When We were Wrong Promptly Admitted it

A personal story in the back of the *Big Book* says, "Nothing, absolutely nothing, happens in God's world by mistake."

To understand the importance of this step, the addict must understand that it is easy to tell the truth when it will cost you nothing, however, it is more difficult when the truth brings about difficult circumstances.

The tenth step states "*continued* to take personal inventory" which implies that this practice had commenced previously. In the *Big Book's* description of the fourth step, it is explained, "resentment is the 'number one' offender. It destroys more alcoholics than anything else...a life which includes deep resentment leads only to futility and unhappiness.... this business of resentment is infinitely grave. We found that it is fatal. For when harboring such feelings we shut ourselves off from the sunlight of the Spirit. The insanity of alcohol returns and we drink again. And with us, to drink is to die. If we were to live, we had to be free of anger" (pp 64–66). The inventory of the fourth step includes a list of the people, principles, and institutions that the individual resents, as well as a fearless and thorough inventory of how the individual was being selfish, dishonest, self-seeking, and fearful—all of which presumably contributed to each specific resentment. Considering that the tendency and opportunity to become resentful toward future people, principles, and institutions is likely inevitable, the *continued* practice of this fearless and thorough inventory is necessary to avoid "deep resentment" and being "shut off from the sunlight of the Spirit." In this way, it is an extension of the fourth and ninth steps, which the creators of the twelve steps continued to practice as a "way of life." As such, the tenth step, along with the eleventh and twelfth steps, is referred to as a *maintenance step*.

According to followers of the twelve-step program—whether the result is good or bad—the person should let his/her truthfulness be known to all so that the individual can live with a clear conscience and be a trustworthy friend. Interestingly, for many in the program, acceptance is the answer to "all my problems today." According to the *Big Book*, "Acceptance is the key to my problems." When an addict is disturbed, it is in many cases because he/she finds some person, place, thing, or situation—some fact of life—unacceptable. Importantly, the person cannot find serenity in daily life until he/she agrees to accept that person, thing, or life situation as being exactly the way it is supposed to be at any moment of time to reiterate: "Nothing, absolutely nothing, happens in God's world by mistake."

Acceptance and truthfulness are not easy especially if the addict is genetically predisposed to lying and manipulation, as was discussed above (e.g., a high association of immature defense style [untruthfulness] in carriers with the DRD2 A1 allele).

This thought brings us to Step 10, which suggests that the addict continues to take personal inventory and continues to set right any new mistakes as they occur. It further suggests that the person vigorously commence this way of living as they cleaned up the past, and that he/she discuss the wrongs with someone immediately and make amends quickly if they have harmed anyone. Accordingly, they have entered the world of the Spirit. The addict continues to watch for selfishness, dishonesty, resentment, and fear. When these crop up, the addict should ask God at once to remove them. This is not an overnight matter. It continues for a lifetime. Moreover, it is suggested.

Step 10 requires that the addict to be vigilant against his/her addictive behaviors and against the triggers of these behaviors. In addition, it requires the person to be humble before God who can keep the person away from addictive behavior if they have the right attitude. It requires them to deal with their defects (genetic and/or environmental) promptly when they arise and not to let them linger in their life (from 12Step.org).

> The emphasis on inventory is heavy only because a great many of addicts have never really acquired the habit of accurate self-appraisal. Working the tenth step becomes a regular part of everyday living, rather than unusual or set apart (*Twelve Steps and Twelve Traditions*, pp 89–90).

In working the tenth step, addicts examine their actions, their reactions, and their motives. Addicts often find that they have been *doing* better than they have been *feeling*. This allows each individual to find out where he/she has gone wrong and admit fault before things get any worse. Most importantly through self-reference addicts need to avoid rationalizing and instead promptly admit their faults, not explain them.

In meditation, both the quality and the contents of consciousness may be voluntarily changed, making it an obvious target in the quest for the neural correlate of consciousness. In a paper on the *mental self*, Lou et al. (2005), from the Department of Functionality Integrative Neuroscience at Aarhus University in

Denmark, described the results of a positron emission tomography study of yoga nidra relaxation meditation when compared with the normal resting conscious state. Meditation was accompanied by a relatively increased perfusion in the sensory imagery system that includes sensory and higher-order association regions and the hippocampus, with decreased perfusion in the executive system: dorsolateral prefrontal cortex, anterior cingulate gyrus, striatum, thalamus, pons, and cerebellum. To identify regions active in both systems, they performed a principal component analysis of the results. This separated the blood flow data into two groups of regions, explaining 25 and 18 % of their variance, respectively: One group corresponded to the executive system and the other to the systems supporting sensory imagery. The inclusion of the striatum, and their subsequent finding of increased striatal DA binding to D2 receptors during meditation, suggested dopaminergic regulation of this circuit.

Lou et al. (2005) then investigated the neural networks supporting episodic retrieval of judgments of individuals with different degrees of self-relevance, in the decreasing order: self, best friend, and the Danish queen. The researchers found that all conditions activated a medial prefrontal–precuneus/posterior cingulate cortex, thalamus, and cerebellum. This activation occurred together with the activation of the left lateral prefrontal/temporal cortex. The latter was dependent on the requirement of retrieval of semantic information, being most pronounced in the "queen" condition. Transcranial magnetic stimulation, targeting precuneus, then was applied to the medial parietal region to transiently disrupt the normal function of the circuitry. They found a decreased efficiency of retrieval of self-judgment compared to the judgment of best friend. This showed that the integrity of the function of precuneus is essential for self-reference, but not for reference to others. This further suggested that self-reference is under control of DA in the brain. Certainly, for carriers of polymorphic genes that compromise DA regulation, this could have profound effects on the ability to garnish the tenth step.

Step 10 has been considered to be a spiritual pocket computer to help keep tabs on behavior today and a cleanser to help keep our spiritual lenses clean. In this method of keeping an inventory every day, the addict must ask questions like:

- Which of my character defects popped up as uninvited guests today?
- Am I using the tools of the program? Am I praying?
- Am I thanking God for all the good things he has done for me this day, and for any positive things he's freed me to do?

Moreover, one's daily inventory certainly needs to assess the status of the addict's relationship with God. Are they still yielding their will to Him? Bill Wilson emphasized how crucial this evaluation is, especially for addictive personalities, which tend to be willful. It has been stated: "…need to surrender ourselves to God on a daily basis will go on throughout our lives, and we shall explore the means of that continuing spiritual surrender in Step 11" (*Serenity, A Companion for Twelve Step Recovery*, pp 67, 69).

A major concern in medical treatment is the concept of compliance, and in some cases, lifelong compliance, whether it is related to the appropriate and continued use of pharmaceuticals, mental therapy, or physical therapy (as with diabetes). This same concept holds true for compliance to the twelve-step program. The reason this is so important is that the *sin-disease*, with its denial and delusion, must be dealt with on a daily basis because it is always hovering "just a decision away" and could throw the addict back into fear and confusion. Its tactics are to convince the addict in various ways, "You're 'well' now and don't need a stupid program to lead a normal life. You can and should operate on your own as a mature adult." Moreover, as Miller (1992) notes in his book, *"A Hunger for Healing,"* when we begin to feel a little secure and happy and relationships are more comfortable, many "forget" to have our quiet time. Many forget to go to meetings and do not call their sponsor. We are busy again, because the pain that drove them into the program has been alleviated. This is a dangerous place to be, because it is one of the major delusions of the spiritual life that addict can "do it ourselves" without daily contact with God and a daily look at the reality of what is going on in our own lives.

This kind of thinking of non-compliance is further problematic when one considers the known hypodopaminergic (genetic) trait coupled with an environmental state that will ultimately throw the individual into relapse, especially in face of not only stress but also anger and fear. Thus, it is noteworthy, that Steps 10, 11, and 12 are sometimes called the maintenance steps. They repeat many of the points outlined in previous steps, but they emphasize the value of taking inventory daily and as such unconsciously raise their DA function through increased synaptic release. This understanding is based in-part on the eloquent work of Molinoff's group going back to 1995 (Boundy et al. 1995). In essence, they found that DA agonists (like DA itself) increased D2 receptors when the kidney-transfected cell was exposed chronically. Specifically, studies with radiolabeled antagonists have revealed that both agonists and antagonists induce up-regulation of D2 DA receptors in cells transfected to express D2L or D2S receptors. The regulation induced by agonists, but not antagonists, was synergistic with cAMP analogs. These findings have been extended by using a radiolabeled agonist to investigate agonist- and antagonist-induced regulation of the high affinity state of the D2L DA receptor in transfected HEK 293 cells. Starr and associates (1995) observed that both agonists, DA and quinpirole, increased the density of D2L receptors. Others also found that DA antagonists increased D2S receptors (Boundy et al. 1996).

Thus taking daily inventory and promptly admitting when there is wrongdoing makes one feel good, and good self-reference concomitantly releases brain DA (Lou et al. 2005) which acts as a D2 agonist and promptly begins to increase compromised D2 receptor densities. Our laboratory is exploring the possibility that attendance to twelve-step programs induce DA activation in the caudate-accumbens.

The tenth step can be a pressure-relief valve. Addicts work this step while the day's ups and downs are still fresh in mind. They list what they have done and try not to rationalize their actions. The first thing they must do is stop! Then, they must take the time to allow themselves the privilege of thinking. They work this step continuously. It presents a way of avoiding grief. The individual monitors feelings, emotions, fantasies, and actions. By constantly looking at these things, they may be able to avoid repeating the actions that make them feel bad (*Narcotics Anonymous Basic Text,* Chapter 4 /*Step 10*). Step 10 is the maintenance step for Steps 4 and 5 and encourages the taking of a personal inventory, which, for recovering persons, should be a daily process. It is important that addicts realize that if they do carry a genetic risk, for example, the DRD2 A1 allele among other gene deficits with 30–40 % less D2 receptor density, taking inventory and feeling good about it is a temporarily "dopamine fix." As such, addicts must continue to "work the steps" on a day-to-day basis to replenish DA.

2.11 Step 11: Sought Through Prayer and Meditation to Improve Our Conscious Contact with God, *as We Understood Him*, Praying Only for Knowledge of His Will for Us and the Power to Carry that Out

> Most of us find that we were neither as terrible, nor as wonderful as we supposed (Narcotics Anonymous).

Step 11 suggests prayer and meditation. "We shouldn't be shy in this matter of prayer. Better men than we are using it constantly. It works, if we have the proper attitude and work at it" (*Big Book*, pp. 85–86).

Accordingly, "Step 11 is my continual reality check and compass. It keeps me grounded in the reality that I know has brought me out of my addictive behaviors. It keeps me in a safe place by keeping my conscious contact with God. Through prayer and meditation I maintain this conscious contact with God and continually try to carry out what God leads me to do. In that path, I find the sanity, serenity and joy that I have been seeking" (from 12Step.org).

Moreover,

> those of us who have come to make regular use of prayer would no more do without it than we would refuse air, food or sunshine. When we refuse air, light or food the body suffers. And when we turn away from meditation and prayer, we likewise deprive our minds, our emotions and our intuitions of vitally needed support. As the body can fail its purpose for lack of nourishment, so can the soul. We all need the light of God's reality, the nourishment of His strength, and the atmosphere of His grace. To an amazing extent the facts of A.A. life confirm this ageless truth (*Twelve Steps and Twelve Traditions*, pp. 97–98).

Importantly, "we become willing to let other people be what they are without having to pass judgment on them. The urgency to take care of things isn't there anymore. We couldn't comprehend acceptance in the beginning-now we can" (*Narcotics Anonymous Basic Text,* Chapter 4/*Step 11*).

Step 11 provides daily spiritual maintenance. As recovering persons, they may use support groups and recovery literature as springboards toward spiritual and emotional growth. They will probably reach a level, though, at which they hunger for an even deeper contact and communication with God.

Taking this journey over time, it has been stated:

> ...as we become comfortable with God, we will talk with Him as with a trusted friend. He will be the Person with whom we can conduct our daily inventories of grief and confession issues. And we will begin to sense His answers to our prayers... (*Serenity, A Companion for Twelve Step Recovery*, pp. 72, 73).

It is essential for the addict and especially the clinician that alcoholism and narcotic addiction are illnesses of the spirit in part because alcoholics and addicts are driven by their disease to behave in ways that compromise their personal ethics and values. In addition, alcoholics and addicts commit crimes that, by the way, may be tied also to the genetic trait rather than just the drug itself. To reiterate, Guo et al. (2007) found a positive association between the self-reported serious and violent delinquency and the *TaqI* polymorphism in the DRD2 gene and the 40-bp VNTR in the DAT1 gene. The trajectories of violent delinquency for the DAT1*10R/9R and DAT1*10R/10R genotype are again about twice as high as that for DAT1*9R/9R (LR test, $p = 0.021$, 2 df). These results only apply to males. Neither variant is associated with delinquency among females. These authors also observed poor judgment related to adolescent unprotected sex. Specifically, Daw and Gou (2011) found that three genes influence whether adolescents use contraception. They found that variants in the DA transporter gene DAT1, the DA receptor gene DRD2, and the monoamine oxidase gene MAOA are associated with unprotected sexual intercourse. Consistent with previous analyses of these data, the genotypes DRD2*A1/A2, DRD2*A2/A2, DAT1*9R/10R, and MAOA*2R are associated with higher odds of unprotected sexual intercourse than other genotypes at these loci. The DRD2 associations apply both to men and to women, whereas the other associations apply to women only. These results are robust to controls for population stratification by continental ancestry; they do not vary by contraceptive type; and they are consistent with previous research showing that these genetic variants are associated with higher rates of impulsivity. This may be important when one considers the impact of having children, not to mention sexually transmitted diseases early on. The findings of hypodopaminergic function due to genetic polymorphisms as cited above may have evolutionary effects. That is, in Comings' book, "*Gene Bomb*" (1996), he suggested the idea that in generations to come there will be an increase in the DRD2 A1 allele, with increasingly more people predisposed to Reward Deficiency behaviors—as high as 74 %, using Bayesian theorem modeling. This is further supported by the work of others suggesting that the well-documented relationship between conduct disorder, the behavioral phenotype of impulsivity, and problematic alcohol/drug use among adolescents may be moderated by A1 carrier status of the DRD2 gene (Esposito-Smythers et al. 2009).

It is also known that under the influence, there are moral and ethical lapses that undermine self-esteem and promote isolation, guilt, shame, poor moral judgment (decision making), and hopelessness. A plethora of studies have related executive function (e.g., decision making, impulsivity, problem solving) and neurotransmitter genetics (see Bowirrat et al. 2012). Individuals differ in their tendencies to make decisions that seek positive outcomes or to avoid negative ones. At the neurobiological level, one model suggests that phasic changes in DA support learning to reinforce good decisions via striatal D1 receptors and to avoid maladaptive choices via striatal D2 receptors. D2 receptor–related genetic contributions to probabilistic avoidance in humans is highlighted from recent data showing that particular DRD2 polymorphisms are associated with functional modulation of receptor expression (Zhang et al. 2007). In one study, Frank and Hutchison (2009) found that a promoter polymorphism rs12364283 associated with transcription and D2 receptor density was strongly and selectively predictive of avoidance-based decisions. Two further polymorphisms (rs2283265 and rs1076560) associated with relatively reduced presynaptic relative to postsynaptic D2 receptor expression were predictive of relative impairments in negative compared to positive decisions. These previously undocumented effects of DRD2 polymorphisms were largely independent of those reported previously by the same authors for the C957T polymorphism (rs6277) associated with striatal D2 density. In contrast, effects of the commonly studied Taq1A polymorphism on reinforcement-based decisions were due to indirect association with C957T. Taken together, these findings suggest multiple D2-dependent genetic mechanisms contributing to avoidance. These findings also underscore the potential for the recovering addict to experience difficulty in overcoming continued poor judgment. However, continued meditation, as suggested in Step 11, and resultant dopaminergic activation at postsynaptic sites (Lou et al. 2005) may work in favor of promoting not only better decision making but also an enhanced spirituality.

It has been noted by many in moving through the twelve-step program, that they increasingly are in contact with some*one*—a *person* rather than a philosophical Higher Power. When this change takes place, many often see miracles happening in their lives. After much fear of losing control, they discover insight, wisdom, power, and courage that they did not have at all. At that point, many say, "I surrender, I give up." They begin to communicate with God concerning what is happening to them. And that is when they are ready to receive the help of Step 11.

> Although these changes happen for many people, they do not happen for all. Many work the steps and stay sane in Twelve-Step programs yet somehow miss the whole thing about prayer and meditation. Most of the people who work good programs, however, are connected to God and do use prayer and meditation in some form. They use them as practical ways of learning who God is and what His will for them may be, as well as for learning useful truths about who they are and what to do in order to find happiness, guidance, peace and continued growth. But mostly they pray because they feel gratitude, love and a sense of awe that the One with whom they are in contact is using his power to heal them ("*A Hunger for Healing,*" by Keith Miller, p. 180).

In this regard, it is important to understand the power of both prayer and specifically mediation on a neurobiological level. Kjaer et al. (2002) was the first to report an association between endogenous neurotransmitter release and conscious experience. Using 11C-raclopride PET, they demonstrated increased endogenous DA release in the ventral striatum during yoga nidra meditation. Yoga nidra is characterized by a depressed level of desire for action, associated with decreased blood flow in prefrontal, cerebellar, and subcortical regions, structures thought to be organized in open loops subserving executive control. In the striatum, DA modulates excitatory glutamatergic synapses of the projections from the frontal cortex to striatal neurons, which in turn project back to the frontal cortex via the pallidum and ventral thalamus. During meditation, 11C-raclopride binding in ventral striatum decreased by 7.9 %. This corresponds to a 65 % increase in endogenous DA release. The reduced raclopride binding correlated significantly with a concomitant increase in EEG theta activity, a characteristic feature of meditation. All participants reported a decreased desire for action during meditation, along with heightened sensory imagery. The level of gratification and the depth of relaxation did not differ between attention and meditation conditions. Here, the researchers showed increased striatal DA release during meditation associated with the experience of reduced readiness for action. It was suggested that being in the conscious state of meditation causes a suppression of cortico-striatal glutamatergic transmission with subsequent DA release. This relaxed state and more DA at the synaptic level promotes inner reflection and even greater acceptance of a Higher Power and possibly a way to understand Him. This is innovative in vivo evidence suggesting regulation of conscious states at a synaptic level.

The same group of researchers also showed that the neurotransmitter DA may function as a regulator of subjective confidence of visual perception in the normal brain. Although much is known about the effect of stimulation by neurotransmitters on cognitive and executive functions (Bowirrat et al. 2012), their effects on subjective confidence of perception had not previously been recorded experimentally. In a controlled study of 24 normal, healthy female university students with the DA agonist pergolide given orally, Lou et al. (2011) show that dopaminergic activation increased confidence in seeing rapidly presented words. It also improved performance in a forced-choice word recognition task. These results demonstrated neurotransmitter regulation of subjective conscious experience of perception and provided evidence for a crucial role of DA. The findings suggest that activation of brain DA may assist the addict in achieving enhanced self-esteem/confidence, thereby facilitating them to perceive the written word of the twelve-step fellowship and exert the effort needed. The notion that DA activation may work in spite of the genetic impact of carrying the DRD2 A1 allele with concomitant 30–40 % less DA D2 receptor density is based on the findings of Molinoff's group. To reiterate, this group showed that DA *per se* as well as other powerful D2 agonists can induce a proliferation of D2 receptors when a cell-like human embryonic kidney 293 is provided with these substance over time (Boundy et al. 1995).

Furthermore, we are cognizant that during the course of an addict's recovery, stress continues to be an unfortunate part of life and is the number one reason for relapse. It is well known that stress significantly reduces brain DA (Blum et al. 2008), and there is also the fact that within-family studies by Madrid et al. (2001) showed a significant association between the DRD2 A1 allele and an inability to cope with stress.

Doing the work required in Step 11 continuously through both the meditative and prayer process increases the release of DA at the synaptic level. In addition, working Step 11 on a daily basis will offset the genetically induced "hypodopaminergic brain function" by continued DA release in the synapse. Increased DA will result in a subsequent proliferation of DA D2 receptors even in carriers of the DRD2 A1 allele and other reward gene polymorphisms. The increase in D2 receptors translates to enhanced DA function, which will ultimately promote greater confidence in the recovering addict, enabling a better understanding of the written word of the twelve-step fellowship. This will lead to an anti-stress effect and as such reduce the chance for relapse, especially in dysfunctional and co-dependent families.

2.12 Step 12: Having had a Spiritual Awakening as the Result of These Steps, We Tried to Carry this Message to Alcoholics, and to Practice These Principles in All Our Affairs

There are three distinctive parts to the twelfth step. The first part states, "having had a spiritual awakening as the result of these steps…. " That means, very simply, if you worked the steps, you had a spiritual awakening. It is not necessarily a description of what *will* happen for you, but rather, of what *has* happened for *us*. Moreover, if an addict works the twelve steps, he or she will indeed know a higher power and attain serenity. The next part of the step says, "We tried to carry this message to alcoholics." That is pretty self-explanatory. We carry the message that God can and will bring peace to the most wretched souls. In short, we make ourselves available to our fellows and share our strength, experience, and hope. In fact, it has been said that those who carry the message experience a feeling of great accomplishment and well-being. This fellowship potentially not only effects DA release but it is also conjectured that twelve-step work also activates the known love-bonding substance oxytocin (de Boer et al. 2012). Lastly, the twelfth step says, "and to practice these principles in all our affairs." This refers to lifestyle of principles that adhere to honesty, spirituality, and a keen interest in others, without judgment. Elaborate evangelizing is not required. This step simply requests that each participant should let others know whether his/her life was improved by the twelve-step fellowship in which he/her was involved. This can be done easily, by saying, "The twelve-steps were helpful to me, and they may be helpful to you as well." It is sharing emotions with another.

In terms of neurobiology, it would be quite interesting if there were a brain chemical that might be involved in sharing one's emotions with another. Oxytocin is a neuropeptide that is attracting growing attention from researchers interested in human emotional and social behavior. There is indeed increasing evidence that oxytocin has a calming effect and that it facilitates pair bonding and social interactions. Some of oxytocin's effects are thought to be direct, but it has been suggested that oxytocin also may have indirect effects, mediated by changes in behavior. One potentially relevant behavioral change is an increased propensity for "emotional sharing" as this behavior, like oxytocin, is known to have both calming and bonding effects.

In a study by Lane et al. (2012), 60 healthy young adult men were randomly assigned to receive either intranasal placebo ($n = 30$) or oxytocin ($n = 30$). Participants were then instructed to retrieve a painful memory. Subsequently, oxytocin and placebo participants' willingness to disclose to another person, event-related facts (factual sharing) versus event-related emotions (emotional sharing) was evaluated. Whereas the two groups were equally willing to disclose event-related facts, oxytocin was found to specifically increase the willingness to share event-related emotions.

This study provided evidence that oxytocin increases people's willingness to share their emotions. Importantly, oxytocin did not make people more talkative (word counts were comparable across the two groups) but instead increased the willingness to share the specific component that is responsible for the calming and bonding effects of social sharing: emotions. These findings suggest that oxytocin may shape the form of social sharing so as to maximize its benefits. This might help explain the calming and bonding effects of oxytocin. Most importantly, under normal physiological processes, the gene that controls either oxytocin synthesis or oxytocin receptors if polymorphic would impact the recovering person to adequately reach this last step.

This notion is highlighted by an important study by Gillath et al. (2008). In their study, the researchers examine associations between attachment insecurities and particular genetic polymorphisms related to emotions and social behavior. They found that (1) anxious attachment was associated with a polymorphism of the DRD2 DA receptor gene, (2) avoidant attachment was associated with a polymorphism of the 5HT2A 5-HT receptor gene, and (3) the rs53576 A polymorphism of the OXTR oxytocin receptor gene was not associated with attachment insecurities. These findings suggested that attachment insecurities are partially explained by particular genes, although there is still a great deal of individual difference variance that remains to be explained by other genes or social experiences.

However, transgenic mice created with the 5′ flanking region of the prairie vole oxytocin receptor gene demonstrated that sequencing in this region influenced the pattern of expression within the brain. The unique promoter sequences of the prairie vole OTR and V1a receptor genes and the resulting species-specific pattern of regional expression provide a potential molecular mechanism for the evolution of pair bonding (love) behaviors and a cellular basis for monogamy (Insel et al. 1998). While this process occurs under normal physiological conditions, it is significantly impaired under the influence of alcohol and possibly by specific drugs of abuse.

According to Silva et al. (2002), data revealed that numerous neurons in the supraoptic nucleus degenerated after prolonged ethanol exposure, and that the surviving neurons increased their activity in order to prevent dramatic changes in water metabolism. Conversely, excess alcohol did not induce cell death in the suprachiasmatic nucleus, but led to depression of neuropeptide synthesis that was further aggravated by withdrawal. Silva et al. (2002) characterized the effects of prolonged ethanol exposure on the magnocellular neurons of the paraventricular nucleus in order to establish whether or not magnocellular neurons display a common pattern of reaction to excess alcohol, irrespective of the hypothalamic cell group to which they belong. Using conventional histological techniques, immunohistochemistry, and in situ hybridization, the structural organization and the synthesis and expression of vasopressin and oxytocin in the magnocellular component of the paraventricular nucleus were studied under normal conditions and following chronic ethanol treatment (6 or 10 months) and withdrawal (4 months after 6 months of alcohol intake). They found that after ethanol treatment, there was a marked decrease in the number of vasopressin- and oxytocin-immunoreactive magnocellular neurons that was attributable to cell death.

Interestingly, opioids are known to reduce neuronal release of oxytocin. In fact, oxytocin release from neurohypophysial terminals is particularly sensitive to inhibition by the micro-opioid receptor agonist, DAMGO. Because the R-type component of the neurohypophysial terminal Ca2 + current (ICa) mediates exclusively oxytocin release, Ortiz-Miranda et al. (2005) hypothesized that micro-opioids could preferentially inhibit oxytocin release by blocking this channel sub-type. Thus, micro-opioid agonists modulate specifically oxytocin release in neurohypophysial terminals by selectively targeting R-type Ca2 + channels. Modulation of Ca2 + channel sub-types could be a general mechanism for drugs of abuse to regulate the release of specific neurotransmitters at central nervous system synapses.

Thus, it is reasonable to assume from a neurobiological point of view that people experience positive feelings when they reach out to others sharing in their emotions, which may, in-part, be due to enhanced release of oxytocin. On the other hand, the negative effects of alcohol and opioids (as examples) prevent active addicts from achieving this state of well-being so important in Step 12 of the fellowship.

Certainly, a major part of Step 12 is the concept of spiritual awakening. References linking genes to complex human traits, such as personality type or disease susceptibility, can be found in the news media and popular culture. In his book, *"The God Gene: How Faith is Hardwired into Our Genes,"* Dean Hamer (*The God Gene: How Faith is Hardwired into Our Genes* (2004) argued that a variation in the VMAT2 gene plays a role in one's openness to spirituality. To reiterate along these lines, Swedish scientists (Nilsson et al. 2007) found that among boys, those with the short 5-HTTLPR genotype showed lower scores, whereas those with the short AP-2beta genotype showed higher scores of the personality character, Self-Transcendence, and its sub-scale Spiritual Acceptance. Among boys and girls, significant interactive effects were found between 5-HTTLPR and AP-2beta genotypes, with regard to Self-Transcendence and Spiritual Acceptance. Boys and girls with the combination of the short 5-HTTLPR, and homozygosity for the long AP-2beta genotype, scored significantly lower on Self-Transcendence and Spiritual Acceptance. This study further supports the view that Spiritual Acceptance to some degree is regulated by our genome. Specifically, carriers of the short 5-HTTLPR and homozygosity for the long AP-2beta genotype will have more difficulty in their ability to experience spiritual awakening and reach Step 12. This fact in no way should discourage anyone from attaining this worthy goal, but clinicians should be aware of this potential pitfall.

Finally, Step 12 refers to lifestyle of principle that adheres to honesty, spirituality, and a keen interest in others without judgment. The recovering addict after achieving the knowledge and necessary work involved in Step 12 should carry these principles for life. However, this may not be achieved easily, and it is impacted both by lifetime traumatic events and by polymorphic genes. Using information pertaining to, for example, pathological gamblers, the effect of genes and environment has been systematically evaluated on lifetime affairs. In common with other addictions, problem and pathological gambling is associated with numerous impairments in the quality of life, including financial, family, legal, and social problems. Gambling disorders (Shapira et al. 2002) commonly co-occur with other psychiatric disorders, such as alcoholism and depression. Scherrer et al. (2005) evaluated male twin members of the Vietnam Era Twin Registry: 53 pathological gamblers, 270 subclinical problem gamblers, and 1346 non-problem gamblers (controls). Results from adjusted logistic regression analyses indicated that, for each mental health domain, pathological gamblers had lower health-related quality of life (HRQoL) scores than problem gamblers ($p < .05$), who in turn had lower scores than non-problem gamblers ($p < .05$). After controlling for genes and family environment, no significant differences existed between the non-problem gambling twins and their problem or pathological gambling brothers, but adjusted co-twin analyses resulted in statistically significant differences in four of eight sub-scales. The authors concluded that pathological and problem gambling is associated with significant decrements in HRQoL. This association was partly explained by genetic and family environmental effects and by lifetime co-occurring substance use disorders. Understanding this result should provide clinicians with some insight into the impact of genes and environment on the ability of the recovering addict to successfully meet the requirements espoused in the last and final fellowship step.

Step 12 occurs when the recovering person had done the work and truly understands all the preceding steps in the program. It has been said that working all the steps will allow an individual to have spiritual awakening. We point out that for people who have addiction, this is dependent on both genes and environmental conditions and attaining spiritual awakening may be more or less difficult. One of the most fulfilling experiences one could get is sharing emotions with others especially as it relates to carrying the message of the fellowship to other addicts. It is important to realize this experience maybe impacted by the synthesis and release of the brain chemical oxytocin. Unfortunately, independent of one's genetic makeup, alcohol and opiates significantly impair the synthesis and release of this important human bonding neuropeptide. Finally, clinicians should be cognizant that any lifestyle change is significantly impacted by both polymorphic genes and traumatic events.

References

Alho H, Miyata M, Korpi E, Kiianmaa K, Guidotti A (1987) Studies of a brain polypeptide functioning as a putative endogenous ligand to benzodiazepine recognition sites in rats selectively bred for alcohol related behavior. Alcohol Alcoholism 1:637–641

Alonso SJ, Navarro E, Rodriguez M (1994) Permanent dopaminergic alterations in the n. accumbens after prenatal stress. Pharmacol Biochem Behav 49(2):353–358

Andrews G (1991) The changing nature of psychiatry. Aust N Z J Psychiatry 25(4):453–459

Andrews G, Pollock C, Stewart G (1989) The determination of defense style by questionnaire. Arch Gen Psychiatry 46(5):455–460

Anscombe GEM (1958) Modern moral philosophy. Philosophy 33:1–19

Aristotle (1996) Introductory Readings. Hackett Pub Co Inc, Cambridge. ISBN 9 9780872203396

Arnsten AF (2009) Toward a new understanding of attention-deficit hyperactivity disorder pathophysiology: an important role for prefrontal cortex dysfunction. CNS Drugs 23(Suppl 1):33–41. doi:10.2165/00023210-200923000-00005

Asghari V, Sanyal S, Buchwaldt S, Paterson A, Jovanovic V, Van Tol HH (1995) Modulation of intracellular cyclic AMP levels by different human dopamine D4 receptor variants. J Neurochem 65(3):1157–1165

Barbaccia ML, Costa E, Ferrero P, Guidotti A, Roy A, Sunderland T, Pickar D, Paul SM, Goodwin FK (1986) Diazepam-binding inhibitor. A brain neuropeptide present in human spinal fluid: studies in depression, schizophrenia, and Alzheimer's disease. Arch Gen Psychiatry 43(12):1143–1147

Barnes JJ, Dean AJ, Nandam LS, O'Connell RG, Bellgrove MA (2011) The molecular genetics of executive function: role of monoamine system genes. Biol Psychiatry 69(12):e127–e143

Bell RP, Foxe JJ, Nierenberg J, Hoptman MJ, Garavan H (2011) Assessing white matter integrity as a function of abstinence duration in former cocaine-dependent individuals. Drug Alcohol Depend 114(2–3):159–168

Bertolino A, Taurisano P, Pisciotta NM, Blasi G, Fazio L, Romano R, Gelao B, Lo Bianco L, Lozupone M, Di Giorgio A, Caforio G, Sambataro F, Niccoli-Asabella A, Papp A, Ursini G, Sinibaldi L, Popolizio T, Sadee W, Rubini G (2010) Genetically determined measures of striatal D2 signaling predict prefrontal activity during working memory performance. PLoS One 5(2):e9348

Blackburn JR, Pfaus JG, Phillips AG (1992) Dopamine functions in appetitive and defensive behaviours. Prog Neurobiol 39(3):247–279

Blanchard MM, Chamberlain SR, Roiser J, Robbins TW, Müller U (2011) Effects of two dopamine-modulating genes (DAT1 9/10 and COMT Val/Met) on n-back working memory performance in healthy volunteers. Psychol Med 41(3):611–618

Blum K, Gold MS (2011) Neuro-chemical activation of brain reward meso-limbic circuitry is associated with relapse prevention and drug hunger: a hypothesis. Med Hypotheses 76(4):576–584

Blum K, Kozlowski GP (1990) Ethanol and neuromodulator interactions: A cascade model of reward. In: Ollat H, Parvaz S, Parvaz H (eds) Alcohol and Behaviour. Prog. Alcohol Res. 2:131–149, VSP, Utecht and Netherlands

Blum K, Topel H (1986) Opioid peptides and alcoholism: genetic deficiency and chemical management. Funct Neurol 1(1):71–83

Blum K, Briggs AH, Elston SF, DeLallo L, Sheridan PJ, Sar M (1982a) Reduced leucine-enkephalin–like immunoreactive substance in hamster basal ganglia after long-term ethanol exposure. Science 216(4553):1425–1427

Blum K, Briggs AH, DeLallo L, Elston SF, Ochoa R (1982b) Whole brain methionine-enkephalin of ethanol-avoiding and ethanol-preferring c57BL mice. Experentia 38(12):1469–1470

Blum K, DeLallo L, Briggs AH, Hamilton MG (1982c) Opioid responses of isoquinoline alkaloids (TIQs). Prog Clin Biol Res 90:387–398

Blum K, Elston SF, DeLallo L, Briggs AH, Wallace JE (1983) Ethanol acceptance as a function of genotype amounts of brain [Met] enkephalin. Proc Natl Acad Sci U S A. 80(21):6510–6512
Blum K, Wallace JE, Briggs AH, Trachtenberg MC (1986) Evidence for the importance of the "genotype" theory in alcohol seeking behavior: a commentary. Alcohol Drug Res 6(6):455–461
Blum K, Briggs AH, Trachtenberg MC, Delallo L, Wallace JE (1987a) Enkephalinase inhibition: regulation of ethanol intake in genetically predisposed mice. Alcohol 4(6):449–456
Blum K, Briggs AH, Wallace JE, Hall CW, Trachtenberg MA (1987b) Regional brain [Met]-enkephalin in alcohol-preferring and non-alcohol-preferring inbred strains of mice. Experientia 43(4):408–410
Blum K, Noble EP, Sheridan PJ, Montgomery A, Ritchie T, Jagadeeswaran P, Nogami H, Briggs AT, Cohn JB (1990) Allelic association of human dopamine D2 receptor gene in alcoholism. JAMA 263:2055–2060
Blum K, Noble EP, Sheridan PJ, Finley O, Montgomery A, Ritchie T, Ozkaragoz T, Fitch RJ, Sadlack F, Sheffield D et al (1991) Association of the A1 allele of the D2 dopamine receptor gene with severe alcoholism. Alcohol 8(5):409–416
Blum K, Braverman ER, Wood RC, Gill J, Li C, Chen TJ, Taub M, Montgomery AR, Sheridan PJ, Cull JG (1996a) Increased prevalence of the Taq I A1 allele of the dopamine receptor gene (DRD2) in obesity with comorbid substance use disorder: a preliminary report. Pharmacogenetics 6(4):297–305
Blum K, Sheridan PJ, Wood RC, Braverman ER, Chen TJ, Cull JG, Comings DE (1996b) The D2 dopamine receptor gene as a determinant of reward deficiency syndrome. J R Soc Med 89(7):396–400
Blum K, Braverman ER, Wu S, Cull JG, Chen TJ, Gill J, Wood R, Eisenberg A, Sherman M, Davis KR, Matthews D, Fischer L, Schnautz N, Walsh W, Pontius AA, Zedar M, Kaats G, Comings DE (1997) Association of polymorphisms of dopamine D2 receptor (DRD2), and dopamine transporter (DAT1) genes with schizoid/avoidant behaviors (SAB). Mol Psychiatry 2(3):239–246
Blum K, Braverman ER, Holder JM, Lubar JF, Monastra VJ, Miller D, Lubar JO, Chen TJ, Comings DE (2000) Reward deficiency syndrome: a biogenetic model for the diagnosis and treatment of impulsive, addictive, and compulsive behaviors. J Psychoactive Drugs 32(Suppl:i-iv):1–112
Blum K, Chen TJ, Meshkin B, Waite RL, Downs BW, Blum SH, Mengucci JF, Arcuri V, Braverman ER, Palomo T (2007) Manipulation of catechol-O-methyl-transferase (COMT) activity to influence the attenuation of substance seeking behavior, a subtype of Reward Deficiency Syndrome (RDS), is dependent upon gene polymorphisms: a hypothesis. Med Hypotheses 69(5):1054–1060
Blum K, Chen AL, Chen TJ, Braverman ER, Reinking J, Blum SH, Cassel K, Downs BW, Waite RL, Williams L, Prihoda TJ, Kerner MM, Palomo T, Comings DE, Tung H, Rhoades P, Oscar-Berman M (2008) Activation instead of blocking mesolimbic dopaminergic reward circuitry is a preferred modality in the long term treatment of reward deficiency syndrome (RDS): a commentary. Theor Biol Med Model 12(5):24. doi:10.1186/1742-4682-5-24
Blum K, Chen ALC, Chen TJH et al (2009a) Genes and Happiness. Gen Ther Mol Biol 13:92–129
Blum K, Chen TJ, Downs BW, Bowirrat A, Waite RL, Braverman ER, Madigan M, Oscar-Berman M, DiNubile N, Stice E, Giordano J, Morse S, Gold M (2009b) Neurogenetics of dopaminergic receptor supersensitivity in activation of brain reward circuitry and relapse: proposing "deprivation-amplification relapse therapy" (DART). Postgrad Med 121(6):176–196
Blum K, Chen TJ, Morse S, Giordano J, Chen AL, Thompson J, Allen C, Smolen A, Lubar J, Stice E, Downs BW, Waite RL, Madigan MA, Kerner M, Fornari F, Braverman ER (2010) Overcoming qEEG abnormalities and reward gene deficits during protracted abstinence in male psychostimulant and polydrug abusers utilizing putative dopamine agonist therapy: part 2. Postgrad med 122(6):214–226. doi:10.3810/pgm.2010.11.2237

Blum K, Chen AL, Oscar-Berman M, Chen TJ, Lubar J, White N, Lubar J, Bowirrat A, Braverman E, Schoolfield J, Waite RL, Downs BW, Madigan M, Comings DE, Davis C, Kerner MM, Knopf J, Palomo T, Giordano JJ, Morse SA, Fornari F, Barh D, Femino J, Bailey JA (2011a) Generational association studies of dopaminergic genes in reward deficiency syndrome (RDS) subjects: selecting appropriate phenotypes for reward dependence behaviors. Int J Environ Res Public Health 8(12):4425–4459

Blum K, Miller M, Perrine K, Liu Y, Giordano J, Oscar-Berman M (2011b) Mesolimbic hypodopaminergic function a potential nutrigenomic therapeutic target for drug craving and relapse. Abstract #pp-248 September 2011, XIX World Congress of Psychiatric Genetics, Washington, DC

Blum K, Bailey JA, Gonzalez AM, Oscar-Berman M, Liu Y, Giordano J, Braverman E, Gold M (2011c) Neurogenetics of reward deficiency syndrome (RDS) as the root cause of "addiction transfer": a new phenomenon common after bariatric surgery. Genet Syndr Gene Ther 2012(1). doi:pii:S2-001

Blum K, Oscar-Berman M, Stuller E, Miller D, Giordano J, Morse S, McCormick L, Downs WB, Waite RL, Barh D, Neal D, Braverman ER, Lohmann R, Borsten J, Hauser M, Han D, Liu Y, Helman M, Simpatico T (2012a) Neurogenetics and nutrigenomics of neuro-nutrient therapy for reward deficiency syndrome (RDS): clinical ramifications as a function of molecular neurobiological mechanisms. J Addict Res Ther 3:139. doi:10.4172/2155-6105.1000139

Blum K, Giordano J, Oscar-Berman M, Bowirrat A, Simpatico T, Barh D (2012b) Diagnosis and healing in veterans suspected of suffering from post-traumatic stress disorder (PTSD) using reward gene testing and reward circuitry natural dopaminergic activation. J Genet Syndr Gene Ther 3:1000116

Blum K, Oscar-Berman M, Barh D, Giordano J, Gold M (2013) Dopamine genetics and function in food and substance abuse. Genet Syndr Gene Ther 4(121). doi:pii:1000121

Bond MP, Vaillant JS (1986) An empirical study of the relationship between diagnosis and defense style. Arch Gen Psychiatry 43(3):285–288

Bond M, Gardner ST, Christian J, Sigal JJ (1983) Empirical study of self-rated defense styles. Arch Gen Psychiatry 40(3):333–338

Borg J, Andrée B, Soderstrom H, Farde L (2003) The serotonin system and spiritual experiences. Am J Psychiatry 11:1965–1969

Boundy VA, Pacheco MA, Guan W, Molinoff PB (1995) Agonists and antagonists differentially regulate the high affinity state of the D2L receptor in human embryonic kidney 293 cells. Mol Pharmacol 48(5):956–964)

Boundy VA, Lu L, Molinoff PB (1996) Differential coupling of rat D2 dopamine receptor isoforms were expressed in Spodoptera frugiperda moth caterpillar cells. J Pharmacol Exp Ther 276(2):784–794

Bousman CA, Cherner M, Atkinson JH, Heaton RK, Grant I, Everall IP (2010) Hnrc group T. COMT Val158Met polymorphism, executive dysfunction, and sexual risk behavior in the context of HIV infection and methamphetamine dependence. Interdiscip Perspect Infect Dis 2010:9, 678648

Bowirrat A, Oscar-Berman M (2005) Relationship between dopaminergic neurotransmission, alcoholism, and Reward Deficiency syndrome. Am J Med Genet B Neuropsychiatr Genet 132B(1):29–37

Bowirrat A, Chen TJ, Oscar-Berman M, Madigan M, Chen AL, Bailey JA, Braverman ER, Kerner M, Giordano J, Morse S, Downs BW, Waite RL, Fornari F, Armaly Z, Blum K (2012) Neuropsychopharmacology and neurogenetic aspects of executive functioning: should reward gene polymorphisms constitute a diagnostic tool to identify individuals at risk for impaired judgment? Mol Neurobiol 45(2):298–313

Braet W, Johnson KA, Tobin CT, Acheson R, McDonnell C, Hawi Z, Barry E, Mulligan A, Gill M, Bellgrove MA, Robertson IH, Garavan H (2011) fMRI activation during response inhibition and error processing: the role of the DAT1 gene in typically developing adolescents and those diagnosed with ADHD. Neuropsychologi 49(7):1641–1650

Braverman ER (1987) The religious medical model: holy medicine and the spiritual behavior inventory. South Med J. 80(4):415–20, 425
Brennan J, Andrews G, Morris-Yates A, Pollock C (1990) An examination of defense style in parents who abuse children. J Nerv Ment Dis 178(9):592–595
Brickman P, Campbell DT (1971) Hedonic relativism and planning the good society. In: Apley MH (ed) Adaptation level theory: a symposium. Academic Press, New York, pp 287–302
Broderick PA, Barr GA, Sharpless NS, Brodger WH (1973) Biogenic amine alterations in limbic brain regions of muricidal rats. Res Comm Chem Pathol Pharmacol 48:3–15
Bruijnzeel AW, Repetto M, Gold MS (2004) Neurobiological mechanisms in addictive and psychiatric disorders. Psychiatr Clin North Am 27(4):661–674
Bunzow JR, Van Tol HH, Grandy DK, Albert P, Salon J, Christie M, Machida CA, Neve KA, Civelli O (1988) Cloning and expression of a rat D2 dopamine receptor cDNA. Nature 336(6201):783–787
Burt A (2009) A mechanistic explanation of popularity: Genes, rule breaking, and evocative gene-environment correlations. J Pers Soc Psychol 96:783–794
Campbell V, Bond R (1982) Evaluation of a character education curriculum. In: McClelland D (ed) Education for Values. Irvington Publishers, New York
Campbell JC, Szumlinski KK, Kippin TE (2009) Contribution of early environmental stress to alcoholism vulnerability. Alcohol 43(7):547–554
Carioppo JT, Patrick W (2008) Loneliness: human nature and the need for social connection. W. W. Norton, New York
Charlton BG (2008) Genospirituality: genetic engineering for spiritual and religious enhancement. Med Hypotheses 71(6):825–828
Chen TJ, Blum K, Mathews D, Fisher L, Schnautz N, Braverman ER, Schoolfield J, Downs BW, Comings DE (2005) Are dopaminergic genes involved in a predisposition to pathological aggression? Hypothesizing the importance of "super normal controls" in psychiatric genetic research of complex behavioral disorders. Med Hypotheses 65:703–707
Chen TJ, Blum K, Chen AL, Bowirrat A, Downs WB, Madigan MA, Waite RL, Bailey JA, Kerner M, Yeldandi S, Majmundar N, Giordano J, Morse S, Miller D, Fornari F, Braverman ER (2011) Neurogenetics and clinical evidence for the putative activation of the brain reward circuitry by a neuroadaptagen: proposing an addiction candidate gene panel map. J Psychoactive Drugs 43(2):108–127
Chopra D (2006) Life after Death. Three Rivers Press, New York
Christakis NA, Fowler JH (2007) The spread of obesity in a large social network over 32 years. N Engl J Med 357(4):370–379 Epub 2007 Jul 25
Cloninger CR, Svrakic DM, Przybeck TR (1993) A psychobiological model of temperament and character. Arch Gen Psychiatry 50(12):975–990
Comings DE (1994a) Candidate genes and association studies in psychiatry. Am J Med Genet 54(4):324–325
Comings DE (1994b) Tourette syndrome: a hereditary neuropsychiatric spectrum disorder. Ann Clin Psychiatry 6(4):235–247
Comings DE (1996) The gene bomb: does higher education and advanced technology accelerate the selection of genes for learning disorders, Addictive and disruptive behaviors?. Hope Press, Durate
Comings DE (2008) Did Man create God? Is your spiritual brain in peace with your thinking brain?. Hope Press, Durate
Comings DE, Blum K (2000) Reward deficiency syndrome: genetic aspects of behavioral disorders. Prog Brain Res 126:325–341
Comings DE, Comings BG, Muhleman D, Dietz G, Shahbahrami B, Tast D, Knell E, Kocsis P, Baumgarten R, Kovacs BW et al (1991) The dopamine D2 receptor locus as a modifying gene in neuropsychiatric disorders. JAMA 266(13):1793–1800
Comings DE, Flanagan SD, Dietz G, Muhleman D, Knell E, Gysin R (1993) The dopamine D2 receptor (DRD2) as a major gene in obesity and height. Biochem Med Metab Biol 50(2):176–185

Comings DE, Muhleman D, Ahn C, Gysin R, Flanagan SD (1994) The dopamine D2 receptor gene: a genetic risk factor in substance abuse. Drug Alcohol Depend 34(3):175–180
Comings DE, MacMurray J, Johnson P, Dietz G, Muhleman D (1995) Dopamine D2 receptor gene (DRD2) haplotypes and the defense style questionnaire in substance abuse, Tourette syndrome, and controls. Biol Psychiatry 37(11):798–805
Comings DE, Rosenthal RJ, Lesieur HR, Rugle LJ, Muhleman D, Chiu C, Dietz G, Gade R (1996a) A study of the dopamine D2 receptor gene in pathological gambling. Pharmacogenetics 6(3):223–234
Comings DE, Ferry L, Bradshaw-Robinson S, Burchette R, Chiu C, Muhleman D (1996b) The dopamine D2 receptor (DRD2) gene: a genetic risk factor in smoking. Pharmacogenetics 6(1):73–79
Comings DE, Gade-Andavolu R, Gonzalez N, Wu S, Muhleman D, Blake H, Mann MB, Dietz G, Saucier G, MacMurray JP (2000a) A multivariate analysis of 59 candidate genes in personality traits: the temperament and character inventory. Clin Genet 58(5):375–385
Comings DE, Gonzales N, Saucier G, Johnson JP, MacMurray JP (2000b) The DRD4 gene and the spiritual transcendence scale of the character temperament index. Psychiatr Genet 10(4):185–189
Comings DE, MacMurray JP (2000) Molecular heterosis: a review. Mol Genet Metab 71(1–2):19–31
Costa E, Guidotti A (1979) Molecular mechanisms in the receptor action of benzodiazepines. Annu Rev Pharmacol Toxicol 19:531–545
Costa E, Guidotti A (1991) Diazepam binding inhibitor (DBI): a peptide with multiple biological actions. Life Sci 49(5):325–344
Costa E, Guidotti A, Mao CC (1975) Involvement of GABA in the action of benzodiazepine–studies on rat cerebellum. Psychopharmacol Bull 11(4):59–60
Costa E, Corda MG, Guidotti A (1983) On a brain polypeptide functioning as a putative effector for the recognition sites of benzodiazepine and beta-carboline derivatives. Neuropharmacology 22(12B):1481–1492
Daw J, Guo G (2011) The influence of three genes on whether adolescents use contraception, USA 1994–2002. Popul Stud (Camb) 65(3):253–271
de Boer A, van Buel EM, Ter Horst GJ (2012) Love is more than just a kiss: a neurobiological perspective on love and affection. Neuroscience 10(201):114–124
Dean JC, Poremba GA (1983) The alcoholic stigma and the disease concept. Int J Addict 18(5):739–751
Di Chiara G, Imperato A (1988) Drugs abused by humans preferentially increase synaptic dopamine concentrations in the mesolimbic system of freely moving rats. Proc Natl Acad Sci USA 85(14):5274–5278
Diana M (2011) The dopamine hypothesis of drug addiction and its potential therapeutic value. Front Psychiatry 2:64. doi:10.3389/fpsyt.2011.00064
Dick DM, Pagan JL, Holliday C, Viken R, Pulkkinen L, Kaprio J, Rose RJ (2007) Gender differences in friends' influences on adolescent drinking: a genetic epidemiological study. Alcoholism: Clin Exp Res 31(12):2012–2019
Dickinson D, Elvevåg B (2009) Genes, cognition and brain through a COMT lens. Neuroscience 164(1):72–87
Diener E, Lucas RE, Scollon CN (2006) Beyond the hedonic treadmill: revising the adaptation theory of well-being. Am Psychol 61(4):305–314
Doris JM (2000) Lack of Character: personality and moral behavior. Cambridge University Press, Cambridge, p 284. ISBN 0521631165
Drew MR, Simpson EH, Kellendonk C, Herzberg WG, Lipatova O, Fairhurst S, Kandel ER, Malapani C, Balsam PD (2007) Transient overexpression of striatal D2 receptors impairs operant motivation and interval timing. J Neurosci 27(29):7731–7739
Dreyer JK, Herrik KF, Berg RW, Hounsgaard JD (2010) Influence of phasic and tonic dopamine release on receptor activation. J Neurosci 30(42):14273–14283
Dudley SA, File AL (2007) Kin recognition in an annual plant. Biol Lett 3:435–438

Duka T, Wüster M, Herz A (1979) Rapid changes in enkephalin levels in rat striatum and by diazepam. Naunyn Schmiedebergs Arch Pharmacol 309(1):1–5
Edge PJ, Gold MS (2011) Drug withdrawal and hyperphagia: lessons from tobacco and other drugs. Curr Pharm Des 17(12):1173–1179
Esposito-Smythers C, Spirito A, Rizzo C, McGeary JE, Knopik VS (2009) Associations of the DRD2 TaqIA polymorphism with impulsivity and substance use: preliminary results from a clinical sample of adolescents. Pharmacol Biochem Behav 93(3):306–312
Ferrero P, Costa E, Conti-Tronconi B, Guidotti A (1986) A diazepam binding inhibitor (DBI)-like neuropeptide is detected in human brain. Brain Res 399(1):136–142
Ford GG (1996) An existential model for promoting life change: Confronting the disease concept. J Subst Abuse Treat 13(2):151–158
Fowler JH, Christakis NA (2008) Dynamic spread of happiness in a large social network: longitudinal analysis over 20 years in the Framingham Heart Study. BMJ 4(337):a2338. doi:10.1136/bmj.a2338
Fowler JH, Dawes CT, Christakis NA (2009) Model of genetic variation in human social networks. Proc Natl Acad Sci U S A 106(6):1720–1724
Fowler JH, Settle JE, Christakis NA (2011) Correlated genotypes in friendship networks. Proc Natl Acad Sci U S A 108(5):1993–1997
Frank MJ, Hutchison K (2009) Genetic contributions to avoidance-based decisions: striatal D2 receptor polymorphisms. Neuroscience 164(1):131–140
Frank E, Salchner P, Aldag JM, Salomé N, Singewald N, Landgraf R, Wigger A (2006) Genetic predisposition to anxiety-related behavior determines coping style, neuroendocrine responses, and neuronal activation during social defeat. Behav Neurosci 120(1):60–71. doi:10.1037/0735-7044.120.1.60
Fromm E (1955) The Sane Society. Rinehart, New York
Gardner EL (2011) Addiction and brain reward and antireward pathways. Adv Psychosom Med 30:22–60
Gelernter J, Kranzler H, Coccaro E, Siever L, New A, Mulgrew CL (1997) D4 dopamine-receptor (DRD4) alleles and novelty seeking in substance-dependent, personality-disorder, and control subjects. Am J Hum Genet 61(5):1144–1152
Gillath O, Shaver PR, Baek JM, Chun DS (2008) Genetic correlates of adult attachment style. Pers Soc Psychol Bull 34(10):1396–1405
Gold MS (1993b) Opiate addiction and the locus coeruleus. The clinical utility of clonidine, naltrexone, methadone, and buprenorphine. Psychiatr Clin North Am Mar 16(1):61–73
Gold MS, Dackis CA (1984) New insights and treatments: opiate withdrawal and cocaine addiction. Clin Ther 7(1):6–21
Goldstein RZ, Volkow ND (2011) Dysfunction of the prefrontal cortex in addiction: neuroimaging findings and clinical implications. Nat Rev Neurosci 12:652–669
Goldstein RZ, Leskovjan AC, Hoff AL, Hitzemann R, Bashan F, Khalsa SS, Wang GJ, Fowler JS, Volkow ND (2004) Severity of neuropsychological impairment in cocaine and alcohol addiction: association with metabolism in the prefrontal cortex. Neuropsychologia 2(11):1447–1458
Gowan T, Whetstone S, Andic T (2012) Addiction, agency, and the politics of self-control: Doing harm reduction in a heroin users' group. Soc Sci Med 74(8):1251–1260
Grandy DK, Litt M, Allen L, Bunzow JR, Marchionni M, Makam H, Reed L, Magenis RE, Civelli O (1989) The human dopamine D2 receptor gene is located on chromosome 11 at q22–q23 and identifies a TaqI RFLP. Am J Hum Genet 45(5):778–785
Greene W, Gold MS (2012) Section 16: drug Abuse. In: Bope ET, Kellerman RD (eds) Conn's Current Therapy. Saunders, Inc., Philadelphia, pp 946–951
Guidotti A, Forchetti CM, Corda MG, Konkel D, Bennett CD, Costa E (1983) Isolation, characterization, and purification to homogeneity of an endogenous polypeptide with agonistic action on benzodiazepine receptors. Proc Natl Acad Sci U S A 80(11):3531–3535
Guo G (2006) Genetic similarity shared by best friends among adolescents. Twin Res Hum Genet 9:113–121

Guo G, Roettger ME, Shih JC (2007) Contributions of the DAT1 and DRD2 genes to serious and violent delinquency among adolescents and young adults. Hum Genet 121(1):125–136

Hack LM, Kalsi G, Aliev F, Kuo PH, Prescott CA, Patterson DG, Walsh D, Dick DM, Riley BP, Kendler KS (2011) Limited associations of dopamine system genes with alcohol dependence and related traits in the Irish Affected Sib Pair Study of Alcohol Dependence (IASPSAD). Alcohol Clin Exp Res 35(2):376–385. doi:10.1111/j.1530-0277.2010.01353.x

Haeffel GJ, Getchell M, Koposov RA, Yrigollen CM, Deyoung CG, Klinteberg BA, Oreland L, Ruchkin VV, Grigorenko EL (2008) Association between polymorphisms in the dopamine transporter gene and depression: evidence for a gene-environment interaction in a sample of juvenile detainees. Psychol Sci 19(1):62–69. doi:10.1111/j.1467-9280.2008.02047.x

Haller J, Makara GB, Kovacs JL (1996) The effect of alpha 2 adrenoreceptor blockers on aggressive behavior in mice: implications for the actions of adrenorector agents. Psychopharmacology 126:345

Haluk DM, Floresco SB (2009) Ventral striatal dopamine modulation of different forms of behavioral flexibility. Neuropsychopharmacology 34(8):2041–2052

Hamer DH (2004) The God Gene: how faith is hardwired into our Genes. Doubleday, New York

Harris GJ, Jaffin SK, Hodge SM, Kennedy DN, Caviness VS, Marinkovic K, Papadimitriou GM, Makris N, Oscar-Berman M (2008) Frontal white matter and cingulum diffusion tensor imaging deficits in alcoholism. Alcoholism: Clin Exp Res 32(6):1001–1013. doi: 10.1111/j.1530-0277.2008.00661.x

Hauge XY, Grandy DK, Eubanks JH, Evans GA, Civelli O, Litt M (1991) Detection and characterization of additional DNA polymorphisms in the dopamine D2 receptor gene. Genomics 10(3):527–530

Herkov MJ, Nias MF, Gold MS (2013) Addiction and dependence. In: Jennings B (ed) Encyclopedia of bioethics, 4th edn. Macmillan, Farmington Hills

Homiak M (2008) Moral character. In: Zalta EN (ed) The stanford encyclopedia of philosophy (Fall Edition). http://plato.stanford.edu/archives/spr2011/entries/moral-character/

Huitt W (2004) Moral and character development. Educational psychology interactive. Valdosta State University, Valdosta. Retrieved 29 April 2011

Hyman SE (2007) The neurobiology of addiction: implications for voluntary control of behavior. Am J Bioeth 7(1):8–11

Iervolino AC, Pike A, Manke B, Reiss D, Hetherington EM, Plomin R (2002) Genetic and environmental influences in adolescent peer socialization: Evidence from two genetically sensitive designs. Child Dev 73:162–174

Insel TR, Winslow JT, Wang Z, Young LJ (1998) Oxytocin, vasopressin, and the neuroendocrine basis of pair bond formation. Adv Exp Med Biol 449:215–224

Iversen SD, Alpert JE (1982) Functional organization of the dopamine system in normal and abnormal behavior. Adv Neurol 35:69–76

Jackson KJ, McIntosh JM, Brunzell DH, Sanjakdar SS, Damaj MI (2009) The role of alpha6-containing nicotinic acetylcholine receptors in nicotine reward and withdrawal. J Pharmacol Exp Ther 331(2):547–554. doi:10.1124/jpet.109.155457.Epub2009Jul30

James W (1902) The varieties of religious experience. A study in human nature. Green and Co. Longmans, New York

Kahneman D, Krueger AB, Schkade D, Schwarz N, Stone AA (2006) Would you be happier if you were richer? A focusing illusion. Science 312(5782):1908–1910

Kendler KS, Baker JH (2007) Genetic influences on measures of the environment: A systematic review. Psychol Med 37:615–626

Kesby G, Parker G, Barrett E (1991) Personality and coping style as influences on alcohol intake and cigarette smoking during pregnancy. Med J Aust 155(4):229–233

Kienast T, Wrase J, Heinz A (2008) Neurobiology of substance-related addiction: findings of neuroimaging. Fortschritte der Neurologie-Psychiatrie 76(Suppl 1):S68–S76

Kirk KM, Eaves LJ, Martin NG (1999) Self-transcendence as a measure of spirituality in a sample of older Australian twins. Twin Res 2(2):81–87

Kirsch P, Reuter M, Mier D, Lonsdorf T, Stark R, Gallhofer B et al (2006) Imaging gene-substance interactions: the effect of the DRD2 TaqIA polymorphism and the dopamine agonist bromocriptine on the brain activation during the anticipation of reward. Neurosci Lett 405:196–201

Kjaer TW, Bertelsen C, Piccini P, Brooks D, Alving J, Lou HC (2002) Increased dopamine tone during meditation-induced change of consciousness. Brain Res Cogn Brain Res 13(2):255–259

Klavis PW, Brady K (2012) Getting to the core of addiction: hatching the addiction egg. Nat Med 18(4):502–503. doi:10.1038/nm.2726

Koob GF, Le Moal M (2008) Neurobiological mechanisms for opponent motivational processes in addiction. Philos Trans R Soc Lond B Biol Sci 363(1507):3113–3123

Koob GF, Volkow ND (2010) Neurocircuitry of addiction. Neuropsychopharmacology 35(1):217–238

Korzybski A (1933/1994) Science and Sanity: An introduction to non-Aristotelian systems and general semantics, (5th ed). International Society For General Semantics, Concord

Kurtz E (2008) Research on alcoholics anonymous: the historical context. *The collected Ernie Kurtz*. Hindsfoot Foundation Series on Treatment and Recovery, Authors Choice, New York, pp 1–22 [originally published 1999]

Lane A, Luminet O, Rimé B, Gross JJ, de Timary P, Mikolajczak M (2012) Oxytocin increases willingness to socially share one's emotions. Int J Psychol pp 1–6

Lawrence LE (1994) Reality is reality. J Natl Med Assoc 86(6):417–419

Lawson-Yuen A, Saldivar JS, Sommer S, Picker J (2008) Familial deletion within NLGN4 associated with autism and Tourette syndrome. Eur J Hum Genet 16(5):614–618

Linnoila M, Virkkunen M, Scheinin M, Nuutila A, Rimon R, Goodwin FK (1983) Low cerebrospinal fluid 5-hydroxyindoleacetic acid concentration differentiates impulsive from nonimpulsive violent behavior. Life Sci 33:2609–2614

Lorenzi M, Karam JH, McIlroy MB, Forsham PH (1980) Increased growth hormone response to dopamine infusion in insulin-dependent diabetic subjects: indication of possible blood-brain barrier abnormality. J Clin Invest 65(1):146–153

Lou HC, Nowak M, Kjaer TW (2005) The mental self. Prog Brain Res 150:197–204

Lou HC, Skewes JC, Thomsen KR, Overgaard M, Lau HC, Mouridsen K, Roepstorff A (2011) Dopaminergic stimulation enhances confidence and accuracy in seeing rapidly presented words. J Vis 11(2):15. doi: 10.1167/11.2.15

Louilot A, Le Moal M, Simon H (1989) Opposite influences of dopaminergic pathways to the prefrontal cortex or the septum on the dopaminergic transmission in the nucleus accumbens. An in vivo voltammetric study. Neuroscience 29(1):45–56

Lyvers M, Tobias-Webb J (2010) Effects of acute alcohol consumption on executive cognitive functioning in naturalistic settings. Addict Behav 35(11):1021–1028

MacDonald SW, Li SC, Bäckman L (2009) Neural underpinnings of within-person variability in cognitive functioning. Psychol Aging 24(4):792–808

Madrid GA, MacMurray J, Lee JW, Anderson BA, Comings DE (2001) Stress as a mediating factor in the association between the DRD2 TaqI polymorphism and alcoholism. Alcohol 23(2):117–122

Makris N, Oscar-Berman M, Jaffin SK, Hodge SM, Kennedy DN, Caviness VS, Marinkovic K, Breiter HC, Gasic GP, Harris GJ (2008) Decreased volume of the brain reward system in alcoholism. Biol Psychiatry 64(3):192–202. doi:10.1016/j.biopsych.2008.01.018

Mandyam CD, Koob GF (2012) The addicted brain craves new neurons: putative role for adult-born progenitors in promoting recovery. Trends Neurosci 35(4):250–260

Markett SA, Montag C, Reuter M (2010) The association between dopamine DRD2 polymorphisms and working memory capacity is modulated by a functional polymorphism on the nicotinic receptor gene CHRNA4. J Cogn Neurosci 22(9):1944–1954

Martinez D, Orlowska D, Narendran R, Slifstein M, Liu F, Kumar D, Broft A, Van Heertum R, Kleber HD (2010) Dopamine type 2/3 receptor availability in the striatum and social status in human volunteers. Biol Psychiatry 67(3):275–278

McBride WJ, Li TK (1998) Animal models of alcoholism: neurobiology of high alcohol-drinking behavior in rodents. Crit Rev Neurobiol 12(4):339–369
McClearn GE (1972) Genetics as a tool in alcohol research. Ann N Y Acad Sci 25(197):26–31
McLellan AT, Lewis DC, O'Brien CP, Kleber HD (2000) Drug dependence a chronic medical illness: implications for treatment, insurance, and outcomes evaluation. J Am Med Assoc 284(13):1689–1695
McNamara P (2009) The neuroscience of religious experience. Cambridge University Press, Cambridge
Mehrabian A, Blum JS (1996) Temperament and personality as functions of age. Int J Aging Hum Dev 42(4):251–269
Miller JK (1992) A hunger for healing: the twelve steps as a classic model for Christian spiritual growth. HarperCollins, New York
Miller NS, Gold MS (1994) Dissociation of "conscious desire" (craving) from and relapse in alcohol and cocaine dependence. Ann Clin Psychiatry 2:99–106
Miller WB, Rodgers JL (2001) The ontogeny of human bonding systems: evolutionary origins, neural bases, and psychological manifestations. Kluwer, Boston
Miller DK, Bowirrat A, Manka M, Miller M, Stokes S, Manka D, Allen C, Gant C, Downs BW, Smolen A, Stevens E, Yeldandi S, Blum K (2010) Acute intravenous synaptamine complex variant KB220™ "normalizes" neurological dysregulation in patients during protracted abstinence from alcohol and opiates as observed using quantitative electroencephalographic and genetic analysis for reward polymorphisms: part 1, pilot study with 2 case reports. Postgrad Med 122(6):188–213
Moore LG, Zamudio S, Zhuang J, Droma T, Shohet RV (2002) Analysis of the myoglobin gene in Tibetans living at high altitude. High Alt Med Biol 3(1):39–47
Moors A, De Houwer J (2005) Automatic processing of dominance and submissiveness. Exp Psychol 52(4):296–302
Muhlenkamp FL, Lucion A, Vogel WH (1995) Effects of selective serotonergic agonists on aggressive behavior in rats. Pharmacol Biochem Behav 50:671–674
Myung W, Lim SW, Kim J, Lee Y, Song J, Chang KW, Kim DK (2010) Serotonin transporter gene polymorphisms and chronic illness of depression. J Korean Med Sci 25(12):1824–1827 Epub 2010 Nov 24
Newberg A (2009) The brain and the biology of belief: an interview with Andrew Newberg, MD. Interview by Nancy Nachman-Hunt. Adv Mind Body Med 24(1):32–36
Nilsson KW, Damberg M, Ohrvik J, Leppert J, Lindström L, Anckarsäter H, Oreland L (2007) Genes encoding for AP-2beta and the Serotonin Transporter are associated with the Personality Character Spiritual Acceptance. Neurosci Lett 411(3):233–237. Epub 2006 Nov 22
Noble EP (1993) The D2 dopamine receptor gene: a review of association studies in alcoholism. Behav Genet 23(2):119–129
Noble EP, Blum K, Ritchie T, Montgomery A, Sheridan PJ (1991) Allelic association of the D2 dopamine receptor gene with receptor-binding characteristics in alcoholism. Arch Gen Psychiatry 48(7):648–654
Noble EP, Noble RE, Ritchie T, Syndulko K, Bohlman MC, Noble LA, Zhang Y, Sparkes RS, Grandy DK (1994) D2 dopamine receptor gene and obesity. Int J Eat Disord 15(3):205–217
Nyman ES, Loukola A, Varilo T, Taanila A, Hurtig T, Moilanen I, Loo S, McGough JJ, Järvelin MR, Smalley SL, Nelson SF, Peltonen L (2012) Sex-specific influence of DRD2 on ADHD-type temperament in a large population-based birth cohort. Psychiatr Genet 22(4):197–201
O'Hara BF, Smith SS, Bird G, Persico AM, Suarez BK, Cutting GR, Uhl GR (1993) Dopamine D2 receptor RFLPs, haplotypes and their association with substance use in black and Caucasian research volunteers. Hum Hered 43(4):209–218
Ortiz-Miranda S, Dayanithi G, Custer E, Treistman SN, Lemos JR (2005) Micro-opioid receptor preferentially inhibits oxytocin release from neurohypophysial terminals by blocking R-type Ca2 + channels. J Neuroendocrinol 17(9):583–590

Oscar-Berman M, Marinkovic K (2007) Alcohol: effects on neurobehavioral functions and the brain. Neuropsychol Rev 17:239–257. doi:10.1007/s11065-007-9038-6

Panksepp J, Knutson B, Burgdorf J (2002) The role of brain emotional systems in addictions: a neuro-evolutionary perspective and new 'self-report' animal model. Addiction 97(4):459–469

Perry JC, Körner AC (2011) Impulsive phenomena, the impulsive character (der Triebhafte Charakter) and DSM personality disorders. J Pers Disord 25(5):586–606. doi:10.1521/pedi.2011.25.5.586

Persico AM, Bird G, Gabbay FH, Uhl GR (1996) D2 dopamine receptor gene TaqI A1 and B1 restriction fragment length polymorphisms: enhanced frequencies in psychostimulant-preferring polysubstance abusers. BiolPsychiatry 40:776–784

Pervin LA (1960) Rigidity in neurosis and general personality functioning. J Abnorm Soc Psychol 61:389–395

Pervin L (1994) A Critical Analysis of Current Trait Theory. Psychol Inq 5:103–113

Pinto E, Ansseau M (2009) Genetic factors of alcohol-dependence. Encephale 35(5):461–469

Pinto E, Reggers J, Gorwood P, Boni C, Scantamburlo G, Pitchot W, Ansseau M (2009) The TaqI A DRD2 polymorphism in type II alcohol dependence: a marker of age at onset or of a familial disease? Alcohol 43(4):271–275

Pollock C, Andrews G (1989) Defense styles associated with specific anxiety disorders. Am J Psychiatry 146(11):1500–1502

Post SG, Puchalski CM, Larson DB (2000) Physicians and patient spirituality: professional boundaries, competency, and ethics. Ann Intern Med 132(7):578–583

Pritchard JK, Di Rienzo A (2010) Adaptation—not by sweeps alone. Nat Rev Genet 11(10):665–667

Rosenquist JN, Murabito J, Fowler JH, Christakis NA (2010) The spread of alcohol consumption behavior in a large social network. Ann Intern Med 152(7):426–433 W141

Rothman RB, Blough BE, Baumann MH (2007) Dual dopamine/serotonin releasers as potential medications for stimulant and alcohol addictions. AAPS J 9(1):E1–E10

Routtenberg A (1987) The reward system of the brain. Sci Am 239(5):154–164

Røysamb E, Tambs K, Reichborn-Kjennerud T, Neale MC, Harris JR (2003) Happiness and health: environmental and genetic contributions to the relationship between subjective well-being, perceived health, and somatic illness. J Pers Soc Psychol 85(6):1136–1146

Ruiz SM, Oscar-Berman M, Sawyer KS, Valmas M, Urban T, Harris GJ (2013) Drinking history associations with regional white matter volumes in alcoholic men and women. Alcoholism: Clin Exp Res 37(1):110–122. doi: 10.1111/j.1530-0277.2012.01862.x

Sapolsky RM (2005) The influence of social hierarchy on primate health. Science 308(5722):648–652

Scherrer JF, Xian H, Shah KR, Volberg R, Slutske W, Eisen SA (2005) Effect of genes, environment and lifetime co-occurring disorders on health-related quality of life in problem and pathological gamblers. Arch Gen Psychiatry 62(6):677–683

Schulz R, Wüster M, Duka T, Herz A (1980) Acute and chronic ethanol treatment changes endorphin levels in brain and pituitary. Psychopharmacology 68(3):221–227

Shapira NA, Ferguson MA, Frost-Pineda K, Gold MS (2002) Gambling and Problem Gambling Prevalence Among Adolescents in Florida. A 100 page Report to the Florida Council on Compulsive Gambling, Inc

Sharpe LG, Pilotte NS, Shippenberg TS, Goodman CB, London ED (2000) Autoradiographic evidence that prolonged withdrawal from intermittent cocaine reduces mu-opioid receptor expression in limbic regions of the rat brain. Synapse 37(4):292–297

Shultes RE, Hoffman A, Ratsch C (1998) Plants of the Gods. Their sacred healing and hallucinogenic powers. Healing Arts Press, Rochester

Silva SM, Madeira MD, Ruela C, Paula-Barbosa MM (2002) Prolonged alcohol intake leads to irreversible loss of vasopressin and oxytocin neurons in the paraventricular nucleus of the hypothalamus. Brain Res 925(1):76–88

Smith DE (2012) The process addictions and the new ASAM definition of addiction. J Psychoactive Drugs 44(1):1–4

Smith GP, Schneider LH (1988) Relationships between mesolimbic dopamine function and eating behavior. Ann N Y Acad Sci 537:254–261
Smith SS, O'Hara BF, Persico AM, Gorelick DA, Newlin DB, Vlahov D, Solomon L, Pickens R, Uhl GR (1992) Genetic vulnerability to drug abuse. The D2 dopamine receptor Taq I B1 restriction fragment length polymorphism appears more frequently in polysubstance abusers. Arch Gen Psychiatry 49(9):723–727
Starr S, Kozell LB, Neve KA (1995) Drug-induced up-regulation of dopamine D2 receptors on cultured cells. J Neurochem 65(2):569–577
Stelzel C, Basten U, Montag C, Reuter M, Fiebach CJ (2009) Effects of dopamine-related gene–gene interactions on working memory component processes. Eur J Neurosci 29(5):1056–1063
Stroth S, Reinhardt RK, Thöne J, Hille K, Schneider M, Härtel S, Weidemann W, Bös K, Spitzer M (2010) Impact of aerobic exercise training on cognitive functions and affect associated to the COMT polymorphism in young adults. Neurobiol Learn Mem 94(3):364–372
Sulmasy DP (2006) Spiritual issues in the care of dying patients: "... it's okay between me and god". JAMA 20; 296(11):1385–1392
Sundaram SK, Huq AM, Wilson BJ, Chugani HT (2010) Tourette syndrome is associated with recurrent exonic copy number variants. Neurology 74(20):1583–1590
Thanos PK, Michaelides M, Umegaki H, Volkow ND (2008) D2R DNA transfer into the nucleus accumbens attenuates cocaine self-administration in rats. Synapse 62(7):481–486
Turchan J, Lasoń W, Budziszewska B, Przewłocka B (1997) Effects of single and repeated morphine administration on the prodynorphin, proenkephalin and dopamine D2 receptor gene expression in the mouse brain. Neuropeptides 31(1):24–28
Uhl G, Blum K, Noble E, Smith S (1993) Substance abuse vulnerability and D2 receptor genes. Trends Neurosci 16(3):83–88
Vaillant GE, Bond M, Vaillant CO (1986) An empirically validated hierarchy of defense mechanisms. Arch Gen Psychiatry 43(8):786–794
van der Zwaluw CS, Engels RC, Vermulst AA, Franke B, Buitelaar J, Verkes RJ, Scholte RH (2010) Interaction between dopamine D2 receptor genotype and parental rule-setting in adolescent alcohol use: evidence for a gene-parenting interaction. Mol Psychiatry 15(7):727–735
van Kammen DP, Guidotti A, Kelley ME, Gurklis J, Guarneri P, Gilbertson MW, Yao JK, Peters J, Costa E (1993) CSF diazepam binding inhibitor and schizophrenia: clinical and biochemical relationships. Biol Psychiatry 34(8):515–22
Vetulani J (2001) Drug addiction. Part II. Neurobiology of addiction. Pol J Pharmacol 53(4):303–317
Volkow ND, Baler RD (2012b) Neuroscience. To stop or not to stop? Science 335(6068):546–548
Volkow ND, Fowler JS, Wang GJ (2003) The addicted human brain: insights from imaging studies. J Clin Invest 111:1444–1451
Volkow ND, Wang GJ, Begleiter H, Porjesz B, Fowler JS, Telang F, Wong C, Ma Y, Logan J, Goldstein R, Alexoff D, Thanos PK (2006) High levels of dopamine D2 receptors in unaffected members of alcoholic families: possible protective factors. Arch Gen Psychiatry 63:999–1008
Volkow ND, Wang GJ, Fowler JS, Tomasi D (2012) Addiction circuitry in the human brain. Annu Rev Pharmacol Toxicol 10(52):321–336
Wang GJ, Volkow ND, Logan J, Pappas NR, Wong CT, Zhu W, Netusil N, Fowler JS (2001) Brain dopamine and obesity. Lancet 357(9253):354–357
Wang GJ, Geliebter A, Volkow ND, Telang FW, Logan J, Jayne MC, Galanti K, Selig PA, Han H, Zhu W, Wong CT, Fowler JS (2011a) Enhanced striatal dopamine release during food stimulation in binge eating disorder. Obesity (Silver Spring) 19(8):1601–1608. doi: 10.1038/oby.2011.27
Wang X, Zhou X, Liao Y, Tang J, Liu T, Hao W (2011b) Microstructural disruption of white matter in heroin addicts revealed by diffusion tensor imaging: a controlled study. Zhong Nan Da Xue Xue Bao Yi Xue Ban 36(8):728–732

Weil A (1972) The natural mind: A revolutionary approach to the drug problem. Houghton Mifflin Company, Boston
Weil A, Rosen W (1983) From chocolates to morphine: understanding mind active drugs. Houghton Mifflin Company, New York
Weiss RD, O'malley SS, Hosking JD, Locastro JS, Swift R (2008) COMBINE Study Research Group (2008) Do patients with alcohol dependence respond to placebo? Results from the COMBINE Study. J Stud Alcohol Drugs 69(6):878–884
What's for dinner? Brain chemical helps people decide. health. http://health.usnews.com/health-news/family-health/brain-and-behavior/articles/2009/11/12/whats-for-dinner-brain-chemical-helps-people-decide.html, November 12, 2009
Wildman WJ (2011) Religious and spiritual experiences. Cambridge University Press, Cambridge
Wise RA, Rompre PP (1989) Brain dopamine and reward. Annu Rev Psychol 40:191–225
Wojnar M, Brower KJ, Strobbe S, Ilgen M, Matsumoto H, Nowosad I, Sliwerska E, Burmeister M (2009) Association between Val66Met brain-derived neurotrophic factor (BDNF) gene polymorphism and post-treatment relapse in alcohol dependence. Alcohol Clin Exp Res 33(4):693–702
Wüster M, Schulz R, Herz A (1980) Inquiry into endorphinergic feedback mechanisms during the development of opiate tolerance/dependence. Brain Res 189(2):403–411
Xu TX, Sotnikova TD, Liang C, Zhang J, Jung JU, Spealman RD, Gainetdinov RR, Yao WD (2009) Hyperdopaminergic tone erodes prefrontal long-term potential via a D2 receptor-operated protein phosphatase gate. J Neurosci 29(45):14086–14099
Dawkins R (2009) The greatest show on Earth: the evidence for evolution. Free Press, Simon and Schuster, Inc, New York
Yaryura-Tobias JA, Nezoroglu FA, Kaplan S (1995) Self-mutilation, anorexia, and dysmenorrhea in obsessive compulsive disorder. Int J Eat Disord 17:33–38
Yuan Y, Zhu Z, Shi J, Zou Z, Yuan F, Liu Y, Lee TM, Weng X (2009) Gray matter density negatively correlates with duration of heroin use in young lifetime heroin-dependent individuals. Brain Cogn 71:223–228
Zhang Y, Bertolino A, Fazio L, Blasi G, Rampino A, Romano R, Lee M-LT, Xiao T, Papp A, Wang D, Sadée W (2007) Polymorphisms in human dopamine d2 receptor gene affect gene expression, splicing, and neuronal activity during working memory. Proc Natl Acad Sci USA 104(51):20552–20557
Zhang Y, Tian J, Yuan K, Liu P, Zhuo L, Qin W, Zhao L, Liu J, von Deneen KM, Klahr NJ, Gold MS, Liu Y (2011) Distinct resting-state brain activities in heroin-dependent individuals. Brain Res 1402:46–53
Zijlstra F, Booij J, van den Brink W, Franken IH (2008) Striatal dopamine D2 receptor binding and dopamine release during cue-elicited craving in recently abstinent opiate-dependent males. Eur Neuropsychopharmacol 18(4):262–270
Zink CF, Tong Y, Chen Q, Bassett DS, Stein JL, Meyer-Lindenberg A (2008) Know your place: neural processing of social hierarchy in humans. Neuron 58(2):273–283

Chapter 3
Conclusions

This book builds upon the AA descriptive work of Chappel and Dupont (1999) and is the first attempt to interpret each step of the twelve-step fellowship in terms of molecular neurobiology. It has been an arduous task, and we the authors humbly accept any criticisms and apologize for any misunderstandings in terms of the fellowship community. It is our sincere goal to encourage others to further our neuroscience base with additional accurate information regarding these powerful 12 step principles, initially set forth as far back as the 1930s. It is in this spirit that we hope the present writings will be viewed and carefully accessed.

The following remarks are not unique but are well known in the fellowship community. The remarks serve as a concise way to delineate the meaning of each step, especially for those in the scientific community not familiar with the twelve-step fellowship.

The first step of a twelve-step fellowship (and/or program) urges a participant to admit he/she is powerless to control certain behaviors or addictions. Only a greater power, traditionally defined as God, is able to restore the sanity of a person whose life has been sabotaged by such behaviors or addictions, according to the second step of the program. However, Step 3 acknowledges that we each nurture a different conception of God. While the Alcoholics Anonymous program, which originally introduced the twelve-step program in the 1930s, was founded on Christian principles, this program later strove to embrace atheists and agnostics as members.

Although the presence of a higher power is suggested in the twelve steps, members of twelve-step programs are not exempt from accountability for their actions, both past and present. In fact, in the fourth step of the program, each incoming participant is asked to make an unflinching moral inventory of his own character. This step seems to be crucial because most relapse occurs between Step 3 and Step 4. This now becomes an active step and as such may go against an individual nature (genes) and past behavior.

Step 5 and Step 6 require participants to admit to God and others the extent of their shortcomings and to allow God to remove these inadequacies of character.

K. Blum et al., *Molecular Neurobiology of Addiction Recovery*,
SpringerBriefs in Neuroscience, DOI: 10.1007/978-1-4614-7230-8_3,

Much like a colleague, God is respectfully considered within the twelve steps. Step 7 urges participants of the program to kindly ask God to help them experience improvement in their lives. Thus, while God is portrayed as a helpful figure in the twelve steps, each member of a twelve-step program alone, is responsible for seeking to improve his own behavior.

Relapse prevention and true recovery are not single events, but rather a continuous reformation process. Rehabilitation begins when a person learns to accept responsibility for any harm that he/she may have caused to himself/herself or others as a result of his/her compulsions. In Step 8 of the twelve-step process, each participant is prompted to make a list of any persons he/she has harmed in the past. Step 9 instructs that he/she should attempt to repair relationships with these people if at all possible. Importantly, Step 10 advises participants to continue to modify their behavior by means of sincere self-analysis. Step 11 counsels participants to practice prayer and meditation routinely, to better understand how to best channel hidden potential in their lives.

Finally, the twelve-step programs, by their nature, are communal in their focus. They do not simply seek to improve the circumstances of an individual person, but rather those of an entire community. That is why most consider it a fellowship. As a result, participants are advised in Step 12 to share their blessings by tending to the needs of others. Elaborate evangelizing is not required. This step requests that each participant should let others know whether the twelve-step program in which he/she was involved improved his/her life. This can be done by saying, "The twelve-steps were helpful to me, and they may be helpful to you as well."

Although there are detractors (see Maltzman's insightful book 2008), a growing body of evidence supports the claim that the AA and the twelve-step programs work! They work for patients who go to meetings and want recovery rather than symptom relief (Betty Ford Institute Consensus Panel 2007; DuPont et al. 2009). It has been stated by many in the fellowship that successfully working through the twelve steps provides the recovering addict with a "brand new psyche" that will translate to a new and improved life of sobriety and or clean time and acceptance of others without judgment. An important question, through the continuous daily inventory, spiritual awakening, and carrying the message to others, is: Is there indeed neuroplasticity and lasting brain change? Evidence suggests that we can measure brain changes by employing reliable neuroimaging tools such as PET, SPECT, fMRI, and MEG. Can we as scientists show that in spite of genetically induced hypodopaminergic traits (genes) and states (environment), preferential neuronal release of DA and even oxytocin in addicts is induced through attendance at self-help meetings, talking to sponsors, and carrying the message to others in need? Can we show that continuous attendance of AA/NA meetings as well as other self-help programs, even in carriers of the DRD2 A1 allele, induces proliferation of DA D2 receptors? Can we show that holistic (comprehensive) approaches such as hyperbaric oxygen therapy, talk therapy, cognitive behavioral therapy, trauma therapy, and sound and music therapy, trauma-relief therapy (TRT) among other known modalities enhance dopaminergic function? Can we show that by using natural D2 agonists such as KB220Z proliferation of DA D2 receptors occurs and, as such, reduces stress, blocks norepinephrine, reduces cravings, increases decision making, enhances social bonding, reduces immature defense style behavior (lying and manipulating), regulates prefrontal cortex–cingulate gyrus dysregulation, enhances focus, reduces withdrawal symptoms, activates NAc DA, enhances self-esteem and confidence, reduces crime, reduces unprotected sex, and among other things induces spiritual awakening?.

One NA member put it into perspective when she said, "The program is perfect and it does not fail—people will."

According to one of us (JG), the 12th step is truly alcoholics and/or addicts new way of getting high by getting out of self and helping others. Moreover, some of the tenants of spirituality of recovery is

- BE KIND INSTEAD OF RIGHT
- DO YOUR BEST NOT TO LIE, CHEAT OR STEAL
- DON'T HURT YOURSELF OR ANYONE ELSE
- HELP YOURSELF IN A POSITIVE WAY
- ABOVE ALL HELP SOMEONE LESS FORTUNATE THAN YOU
- REMEMBER NOTHING IS IMPOSSIBLE
- BELIEVE UNTIL IT BECOMES BELIEVABLE.

The embracing of a higher power may be different for many individuals and to some it may represent nature and everything in it. One of us (KB) believe the words of the great American architect Frank Lloyd Wright, who bridged the gap by suggesting that if we follow the laws of nature we will intuitively find peace and happiness:

> "LOVE is the virtue of the heart
> SINCERITY the virtue of the mind
> COURAGE the virtue of the spirit
> DECISION the virtue of the will".
> Frank Lloyd Wright: The Organic Commandment 1940.

Finding happiness may not only reside in our genome but may indeed be impacted by positive meditative practices, positive psychology (Powers 2012) *spiritual acceptance, love of others, and taking inventory of ourselves—one day at a time.*

References

Betty Ford Institute Consensus Panel (2007) What is recovery? A working definition from the Betty Ford Institute. J Subst Abuse Treat 33:221–228

Chappel JN, Dupont RL (1999) Twelve-step and mutual-help programs for addictive disorders. Psychiatr Clin N Am 22(2):425–446. doi:10.1016/S0193-953X(05)70085-X PMID 10385942

DuPont RL, McLellan AT, White WL, Merlo LJ, Gold MS (2009) Setting the standard for recovery: physicians' Health programs. J Subst Abuse Treat 36(2):159–171

Maltzman I (2008) Alcoholism: its treatments and mistreatments. World Scientific, New Jersey

Powers J (2012) When the servant becomes the Master. Central Recovery Press, Las Vegas

Wright FL, Noll S (1940) The Organic Commandment 1(1) Taliesin, New York

Index

K. Blum et al., *Molecular Neurobiology of Addiction Recovery*, SpringerBriefs in Neuroscience, DOI: 10.1007/978-1-4614-7230-8,

Made in the USA
San Bernardino, CA
02 September 2015